A Concise Introduction to Additives for Thermoplastic Polymers

Scrivener Publishing
3 Winter Street, Suite 3
Salem, MA 01970

Scrivener Publishing Collections Editors

James E. R. Couper	Richard Erdlac
Rafiq Islam	Pradip Khaladkar
Norman Lieberman	Peter Martin
W. Kent Muhlbauer	Andrew Y. C. Nee
S. A. Sherif	James G. Speight

Publishers at Scrivener
Martin Scrivener (martin@scrivenerpublishing.com)
Phillip Carmical (pcarmical@scrivenerpublishing.com)

A Concise Introduction to Additives for Thermoplastic Polymers

Johannes Karl Fink
Montanuniversität Leoben, Austria

Scrivener

Copyright © 2010 by Scrivener Publishing, LLC. All rights reserved.

Co-published by John Wiley & Sons, Inc., Hoboken, New Jersey and Scrivener Publishing, LLC, Salem, Massachusetts.
Published simultaneously in Canada.

No part of this publication may be reproduced, stored in a retrieval system, or transmitted in any form or by any means, electronic, mechanical, photocopying, recording, scanning, or otherwise, except as permitted under Section 107 or 108 of the 1976 United States Copyright Act, without either the prior written permission of the Publisher, or authorization through payment of the appropriate per-copy fee to the Copyright Clearance Center, Inc., 222 Rosewood Drive, Danvers, MA 01923, (978) 750-8400, fax (978) 750-4470, or on the web at www.copyright.com. Requests to the Publisher for permission should be addressed to the Permissions Department, John Wiley & Sons, Inc., 111 River Street, Hoboken, NJ 07030, (201) 748-6011, fax (201) 748-6008, or online at http://www.wiley.com/go/permission.

Limit of Liability/Disclaimer of Warranty: While the publisher and author have used their best efforts in preparing this book, they make no representations or warranties with respect to the accuracy or completeness of the contents of this book and specifically disclaim any implied warranties of merchantability or fitness for a particular purpose. No warranty may be created or extended by sales representatives or written sales materials. The advice and strategies contained herein may not be suitable for your situation. You should consult with a professional where appropriate. Neither the publisher nor author shall be liable for any loss of profit or any other commercial damages, including but not limited to special, incidental, consequential, or other damages.

For general information on our other products and services or for technical support, please contact our Customer Care Department within the United States at (800) 762-2974, outside the United States at (317) 572-3993 or fax (317) 572-4002.

Wiley also publishes its books in a variety of electronic formats. Some content that appears in print may not be available in electronic format. For information about Wiley products, visit our web site at www.wiley.com.

For more information about Scrivener products please visit www.scrivenerpublishing.com.
Cover designed by Russell Richardson.

Library of Congress Cataloging-in-Publication Data is available.

ISBN 978-0-470-60955-2

Printed in the United States of America.

10 9 8 7 6 5 4 3 2 1

Contents

Preface		xiii
1	**Introduction**	1
	1.1 Classification	1
	References	3
2	**Plasticizers**	5
	2.1 Principle of Action	6
	2.2 Principle of Selection	6
	2.3 Characterization	7
	2.4 Risks and Drawbacks	8
	2.4.1 Leaching	8
	2.4.2 Inherent Toxicity	9
	2.5 Classes of Plasticizers	9
	2.5.1 Phthalate Plasticizers	9
	2.5.2 Cyclohexanoic Diesters	12
	2.5.3 Phophate Plasticizers	13
	2.5.4 Aliphatic Esters	13
	2.5.5 Polymeric Plasticizers	14
	2.5.6 Ionic Liquids	15
	2.6 Specific Examples of Application	16
	2.6.1 Heat Shrinkable Films	16
	2.6.2 Adhesive Compositions	17
	2.6.3 Interlayer Films for Safety Glasses	17
	2.6.4 Electrolyte Membranes	18
	2.6.5 Porous Electrodes	18
	2.6.6 Biodegradable Polymers	19
	2.6.7 Plasticizers for Energetic Polymers	20
	References	21

3 Fillers — 25
- 3.1 Surface Modification — 25
  - 3.1.1 Siloxanes — 25
  - 3.1.2 Dispersion and Coupling Additives — 25
- 3.2 Special Applications — 29
  - 3.2.1 Flame Retardant Fillers — 29
  - 3.2.2 Conductive Fillers — 30
  - 3.2.3 Solder Precoated Fillers — 32
  - 3.2.4 Nano Clays — 33
  - 3.2.5 Mixed Matrix Membranes — 34
- References — 34

4 Colorants — 37
- 4.1 Physics Behind a Color — 37
  - 4.1.1 Human Eye — 37
  - 4.1.2 Tristimulus Values — 37
  - 4.1.3 Color Spaces — 39
- 4.2 Color Index — 40
- 4.3 Test Standards — 41
- 4.4 Pigments — 43
- 4.5 Organic Colorants — 43
- References — 47

5 Optical Brighterners — 49
- 5.1 Basic Principles — 50
- 5.2 Measurement — 51
- 5.3 Inorganic Brighteners — 52
- 5.4 Organic Optical Brighteners — 52
  - 5.4.1 Reactive Optical Brighteners — 52
  - 5.4.2 Melt Extrusion — 54
  - 5.4.3 Photographic Supports — 55
- References — 56

6 Antimicrobial Additives — 59
- 6.1 Modes of Action — 59
  - 6.1.1 Types of Irritations — 60
- 6.2 Plasticizers — 60

	6.3	Special Formulations	65
		6.3.1 Contact Lenses	65
		6.3.2 Food Packaging	66
		6.3.3 Polymers with Inherent Antimicrobial Properties	67
	References	68
7	**Flame Retardants**	**71**
	7.1	Mechanisms of Flame Retardants	71
		7.1.1 Flame Cooling of Halogens	71
	7.2	Smoke Suppressants	73
	7.3	Admixed Additives	74
	7.4	Bonded Additives	77
		7.4.1 Examples of Polymers	77
	References	83
8	**Lubricants**	**87**
	8.1	Principle of Action	87
	8.2	Methods of Incorporation	88
		8.2.1 Conventional Method	88
		8.2.2 Separate Delivery of the Lubricant	88
	8.3	Types of Lubricants	89
		8.3.1 Alcohols	89
		8.3.2 Fatty Acids, Esters and Amides	90
		8.3.3 Waxes	90
		8.3.4 Polymeric Lubricants	91
	8.4	Special Applications	91
		8.4.1 PVC	91
		8.4.2 Chlorinated PVC	91
		8.4.3 Electically Conductive Polymers	92
	References	92
9	**Antistatic Additives**	**95**
	9.1	Types of Additives	95
	9.2	Areas of Application	96
	9.3	Additives in Detail	98
		9.3.1 Conventional Additives	98
		9.3.2 Polymeric Additives	100

9.3.3	External Antistatic Additives	101
9.3.4	Intrinsically Antistatic Compositions	101
9.3.5	Conductive Fillers	101
References		104

10 Slip Agents — 107
10.1 Basic Principles of Action — 107
10.2 Compounds — 109
10.3 Special Formulations — 110
    10.3.1 Poly(ethylene terephthalate) — 110
    10.3.2 Formulations for Poly(ethylene) — 111
References — 111

11 Surface Improvers — 113
11.1 Additives — 114
    11.1.1 Fluorocarbon Compounds — 114
    11.1.2 Acrylics — 114
    11.1.3 Modified Pigments — 115
    11.1.4 Organic Salts — 116
References — 116

12 Nucleating Agents — 119
12.1 Crystalline Polymers — 120
    12.1.1 Crystal Structures — 120
    12.1.2 Modification of Properties by Crystallinity — 120
12.2 Experimental Methods — 121
    12.2.1 Nucleation Technologies — 121
    12.2.2 Characterization of Polymer Crystallization — 121
12.3 Classes of Nucleating Agents — 122
    12.3.1 Inorganic Nucleating Agents — 122
    12.3.2 Sorbitol Compounds — 123
    12.3.3 Phosphates — 124
    12.3.4 Carbon Nanotubes — 124
    12.3.5 Coupled Nucleating Agents — 124

12.4 Crystallization Accelerators ... 124
12.5 Clarifying Agents ... 125
References ... 125

13 Antifogging Additives ... 127
13.1 Field of Use ... 127
13.2 Principles of Action ... 128
    13.2.1 Thermodynamics of Surfaces ... 128
13.3 Conventional Compounds ... 130
13.4 Compounds for Grafting ... 133
References ... 135

14 Antiblocking Additives ... 137
14.1 Examples of Uses ... 138
    14.1.1 Film Resins ... 138
    14.1.2 Sealable Coatings ... 140
    14.1.3 Membranes ... 141
    14.1.4 Poly(vinyl butyral) ... 142
References ... 144

15 Hydrolysis ... 145
15.1 Hydrolytic Degradation ... 145
    15.1.1 Ordinary Hydrolysis ... 145
    15.1.2 Enzymatic Hydrolysis ... 146
    15.1.3 Stabilization ... 146
15.2 Polymers ... 146
    15.2.1 Poly(ester)s ... 147
    15.2.2 Poly(ester urethane)s ... 147
    15.2.3 Poly(lactide)s ... 148
References ... 149

16 Dehydrochlorination Stabilizers ... 151
16.1 Dehydrochlorination of PVC ... 151
16.2 Stabilizers ... 152
    16.2.1 Alkyl Tin Compounds ... 154
    16.2.2 Mixed Metal Compounds ... 155
    16.2.3 β-Diketones ... 155

16.2.4 Epoxidized fatty acids	155
16.2.5 Hydrotalcite Clays	156
16.2.6 Zeolites	157
16.2.7 Costabilizers	158
References	158

17 Acid Scavengers — 161
17.1 Acid Scavenging — 161
17.2 Examples of Formulation — 162
    17.2.1 Poly(olefin)s — 162
    17.2.2 Poly(ethylene terephthalate) — 163
    17.2.3 Poly(urethane)s — 163
References — 163

18 Metal Deactivators — 165
18.1 Action of Metals in Polymers — 165
18.2 Usage — 166
    18.2.1 Residues of Catalysts — 166
    18.2.2 Metallic Reinforcing Parts — 166
18.3 Examples of Metal Deactivators — 167
    18.3.1 Side Effects — 169
References — 170

19 Oxidative Degradation — 173
19.1 Autoxidation — 173
19.2 Inhibition of Autoxidation — 174
References — 183

20 Degradation by Light — 185
20.1 Photolysis — 189
    20.1.1 Redox Catalysis — 190
    20.1.2 Scavenging — 190
20.2 Photooxdation — 191
20.3 UV Stabilizers — 193
References — 195

21 Blowing Agents — 197
21.1 Blowing Agents — 197
    21.1.1 Physical Blowing Agents — 198

	21.1.2 Chemical Blowing Agents	199
21.2	Ozone Depletion Potential	202
21.3	Test Methods	203
21.4	Special Applications	203
	21.4.1 Poly(urethane) Foams	203
	21.4.2 Poly(imide) Foams	205
	21.4.3 Poly(ethylene) Foams	205
	References	206

22 Compatibilizers — 209

22.1	Estimation of Compatibility	210
	22.1.1 Glass Transition Temperature	210
	22.1.2 Hildebrand Solubility Parameters	211
22.2.	Compatibilizers	215
	22.2.1 Classification	217
	22.2.2 Reactive Processing	221
22.3	Special Examples	221
	22.3.1 Block Copolymers as Compatibilizers	221
	22.3.2 Poly(olefin) Blends	222
	22.3.3 Poly(amide) Blends	223
	22.3.4 Poly(carbonate) Blends	224
	22.3.5 Composites of PVC and Cellulosic Materials	225
	22.3.6 Packaging Applications	226
	References	227

23 Prediction of Service Time — 233

23.1	Accelerated Aging	233
	23.1.1 Cumulative Material Damage	233
	23.1.2 Arrhenius Extrapolation	234
	23.1.3 Interference of Phase Transitions	235
23.2	Theory of Critical Distances	235
23.3	Monte Carlo Methods	236
23.4	Issues in Matrix Composites	236
	References	236

24 Safety and Hazards — 239

24.1	Plasticizers	239
	24.1.1 Di(2-ethylhexyl)phthalate	239

xii   Contents

      24.1.2 Ingestion of PVC   240
      24.1.3 Tricresyl Phosphate   242
  24.2 Flame Retardants   242
      24.2.1 HET-Acid   242
      24.2.2 Brominated Diphenyl Ethers   243
  24.3 Antifogging Agents   243
  24.4 Other   244
      24.4.1 Bisphenol A   244
      24.4.2 Azodicarbonamide   244
  References   245

Index   **249**
  Acronyms   249
  Chemicals   251
  General Index   261

Preface

This book focuses on additives for thermoplastic polymers.

There are many excellent books dealing with additives for polymers. They range from the large 1000-page tomes such as *Plastics Additives Handbook* edited by Zweifel and *Plastics Additives and Modifiers Handbook* edited by Edenbaum, down to the books with slightly less weight and pages such as *Additives for Plastics Handbook* by Murphy.

While all these books are aimed at the practitioner, the very size of them deters or hinders the new person to the field who wants to get a comprehensive yet introductory overview of the subject. I have written this book with the purpose of rectifying this problem and I hope you find that I have succeeded.

The idea for this book came out of a course I was teaching on plastics technology. As I prepared the lecture notes I realized there was a shortage of teaching material on additives for thermoplastic polymers so I have tried to fill the gap. The goal of the book is to offer a general and concise introduction into plastics additives. For students who will be engaged later in the development of plastics formulation, the book will serve as a basic introduction and as a stepping stone to the more detailed books. For those who go into polymer science this book will be sufficient as it gives enough understanding of the specialists' needs.

Beyond education, this book will serve the needs of industry engineers and specialists who have only a passing contact with the plastics industry but need to know more.

Johannes Karl Fink

Leoben
September 2009

1
Introduction

There are many excellent monographs dealing with additives for polymers. The most famous is that of Gächter and Müller, recently edited by Hans Zweifel (1, 2). Other books include the book of Murphy and others (3–7).

News and forthcoming events with regard to both additives and the techniques of incorporating them can be found in journals entitled *Plastics, Additives and Compounding* and *Additives for Polymers*.

1.1 Classification

Additives can be classified according to several criteria, i.e.:

- Field of Application,
- Chemical structure,
- Molecular weight,
- Mode of action,
- Polymer type to be used,
- Reactiveness,
- Effectiveness,
- Side effects (multipurpose action), or
- Commercial importance.

The most comprehensive classification is the classification with respect to its field of application. This kind of classification is summarized in Table 1.1.

Additives can be subdivided into chemically inert additives and chemically reactive additives. For example, plasticizers, or lubricants are not chemically reactive. On the other hand, antioxidants

Table 1.1: Classification of Additives for Polymers

Type	Usage for
Antioxidant	Service time
Light stabilizer	Service time
Acid scavenger	PVC
Lubricant	Processing aid
Processing aid	Unspecific
Antiblocking	Packaging
Slip additive	Packaging
Antifogging additive	Greenhouse
Antistatic additive	General Purpose
Antimicrobial agent	Amides, esters, urethanes
Flame retardant	Safety
Blowing agent	Foams
Modifier	Unspecific term
Controlled degradation additive	Reactive molding
Crosslinker	Reactive molding
Colorant	Beauty
Filler	Mechanical
Reinforcement	Mechanical
Optical whitener	Beauty
Coupling agent	Filler matrix coupling
Nucleating agent	Mechanical
Recycling aid	Environmental
Doping agent	Optoelectronics

are not or should not be chemically reactive when incorporated into the polymeric matrix, but they will become chemically reactive when they are starting with their protective action. The same is mostly true for a flame retardant, but this not a general rule.

In addition, there is a basic difference between additives for thermoplastic material and additives for thermosetting resins. Likewise, a curing agent and an accelerator may be considered as an additive. However, these types of additives are not usually considered as additives in the common sense, so they are not taken up into this book.

Moreover, there are additives that can be rarely found in general texts on additives. For example, additives that are used in organic light emitting diodes are usually omitted in the discussion.

Thus, the definition of what is an additive and what is not an additive is somewhat blurry. Furthermore, it does not make sense to search for an airtight definition because such a definition would be highly complicated to build and would be very difficult to understand.

References

1. H. Zweifel, ed., *Plastics Additives Handbook*, Hanser Publishers, Munich, 5th edition, 2001.
2. H. Zweifel, R.D. Maier, and M. Schiller, eds., *Plastics Additives Handbook*, Hanser Publishers, Munich, 6th edition, 2009.
3. H.H.G. Jellinek, ed., *Degradation and Stabilization of Polymers. A Series of Comprehensive Reviews*, Vol. 2, Elsevier, Amsterdam, New York, 1989.
4. J. Murphy, *Additives for Plastics Handbook*, Elsevier Advanced Technology, Oxford, 2nd edition, 2001.
5. T.A. Osswald, *International Plastics Handbook: The Resource for Plastics Engineers*, Carl Hanser Verlag, Munich, Vienna, New York, 2006.
6. J. Edenbaum, ed., *Plastics Additives and Modifiers Handbook*, Chapman & Hall, London, 1996.
7. J.C.J. Bart, *Additives in polymers: Industrial analysis and applications*, John Wiley, New York, 2005.

2
Plasticizers

Plasticizers serve to soften polymeric materials. Thus, the primary role of plasticizers is to improve the flexibility and processability of polymers. This is achieved by lowering the second order transition temperature of the particular polymer (1). The greatest amount of plasticizers goes into poly(vinyl chloride) (PVC).

Plasticizers have long been known for their effectiveness in producing flexible plastics for applications ranging from the automotive industry to medical and consumer products. In the early days, camphor was used to plasticize celluloid. Soon afterwards, camphor was substituted by tricresyl phosphate. This compound is still in use for PVC. Phthalic acid esters were introduced in 1920 and are still the most important class of plasticizers today.

Recent plasticizer research has focused on technological challenges including leaching, migration, evaporation, and degradation of plasticizers, each of which eventually lead to deterioration of thermomechanical properties in plastics (2).

Approaches to reduce evaporation and degradation of plasticizers have been developed, with the aim of formulating long-lasting flexible plastics and minimizing the ultimate environmental impact of these chemicals. Also, fire-retardant plasticizers and plasticizers for use in biodegradable plastics have been developed (2).

Several monographs have been prepared with regard to the topic (3–7). Plasticizers are used for several types of polymers, including:

- Poly(vinyl chloride),
- Acrylics,
- Aminoplasts,
- Cellulosics,

- Epoxy,
- Phenolic,
- Poly(amide),
- Poly(urethane),
- Styrene-butadiene,
- Miscellaneous vinyl resins,
- Linear poly(ester)s, and
- Elastomers.

The most frequently plasticized polymers include PVC, poly(vinyl butyral), poly(vinyl acetate) (PVAc), acrylics, cellulose molding compounds, and poly(amide)s. About 80% of all plasticizers are used in PVC.

2.1 Principle of Action

Plasticizers exhibit usually low molecular weight. They are forming secondary bonds with polymer chains and thus increase the intermolecular distance of the polymer chains. In other words, they spread the polymer chains apart (2).

For this reason, plasticizers reduce the side valence bonding forces of the chains and establish more mobility for the macromolecules. Consequently, a softer, more easily deformable bulk material is obtained.

In crystalline polymers, the crystalline region remains unaffected, because plasticizers enter only the amorphous regions of polymers.

Plasticizers are reducing the modulus, tensile strength, hardness, density, melt viscosity, glass transition temperature, electrostatic chargeability and volume resistivity of a polymer. In contrast, they are increasing the flexibility, elongation at break, toughness, dielectric constant and power factor (2).

In order to avoid phase separation, plasticizers should be highly compatible with the base polymer.

2.2 Principle of Selection

Plasticizers are generally selected on the basis of the following criteria (8, 9):

- Compatibility of a plasticizer with a given polymer,
- Processing characteristics,
- Desired thermal, electrical and mechanical properties of the end product,
- Resistance to water, chemicals, solar radiation, weathering, dirt, microorganisms,
- Effect of plasticizer on rheological properties of polymer,
- Toxicity, and
- Costs.

2.3 Characterization

The durometer test is based on the penetration of a specific type of indentor when forced into the material under specified conditions. The indentation hardness is inversely related to the penetration and is dependent on the elastic modulus and viscoelastic behavior of the material. The geometry of the indentor and the applied force influence the measurements such that no simple relationship exists between the measurements obtained with one type of durometer and those obtained with another type of durometer or other instruments used for measuring hardness (10).

The durometer test is an empirical test intended primarily for control purposes. No simple relationship exists between indentation hardness determined by this test method and any fundamental property of the material tested. Various types of indentors are in use, as shown in Figure 2.1.

Figure 2.1: Durometer Test (10)

The efficiency of a plasticizer is the change of any desired property referred to its weight added (12). The hardness of various epoxidized palmoil esters is shown in Figure 2.2.

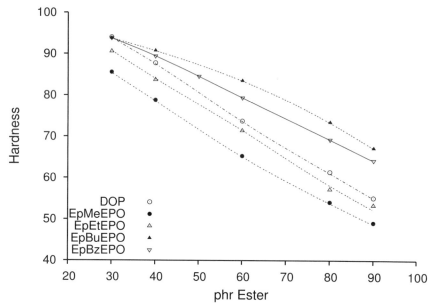

Figure 2.2: Hardness of Various Epoxidized Esters (11)

2.4 Risks and Drawbacks

2.4.1 Leaching

Leaching and migration of plasticizer molecules from polymers is a critical issue that impacts the service time spent on an article.

Leaching, refers to the removal of a substance in a solid material by the extraction of a liquid medium. On the other hand, migration refers to any phenomenon by which a component escapes from a material.

Polymers are often in contact with fluids. Thus, in the course of time, plasticizers may diffuse to the polymer surface and cross over into the external medium. Often, the permeation step has been found to be the limiting step rather than diffusion of plasticizer through the matrix.

In particular, plasticizers escaping from the polymer, often provide toxicity risks for health and environment. Therefore, leaching and migration issues are one of the most important problems in this topic.

Leaching and migration of plasticizers from polymer surface can be reduced by coating the polymer surface.

2.4.2 Inherent Toxicity

Human exposure to certain plasticizers has been debated because di-(2-ethylhexyl) phthalate (DOP), used in medical plastics, has been found at detectable levels in the blood supply and potential health risks may arise from its chronic exposure. A further issue is the use of phthalates in baby-care products and toys. Since young children often put their plastic toys in the mouth, the plasticizers are prone to be leached out and can be swallowed (2). Research with animals revealed a possible endocrine-disruption activity (13).

Benzoate based plasticizers, e.g., Benzoflex® 2888 which is a blend of diethylene glycol dibenzoate, triethylene glycol dibenzoate, and dipropylene glycol dibenzoate, have been developed to account for the leaching problems. Benzoflex® seems to be a good alternative to phthalates in flexible toys due to its ease of processing, final product performance, low toxicity and fast biodegradation. Toxicity tests showed a low acute toxicity and no evidence of reproductive toxicity (2).

2.5 Classes of Plasticizers

Plasticizers can be classified according to their chemical structure as shown in Table 2.1. Plasticizers may be also classified into primary and secondary types (14). Primary plasticizers are used solely as plasticizer, i.e., as the basic component of the plasticizer formulation. Secondary plasticizers are blended with primary plasticizers in order to improve some of the properties.

2.5.1 Phthalate Plasticizers

Phthalate based plasticizers are summarized in Table 2.2. In phthalate esters, the benzene nucleus highly enhances the compatibility to PVC. However, the compatibility decreases with increasing length of the alkyl chains. Phthalates with short alkyl chains are easier to formulate since they diffuse faster. However, a drawback is that

Table 2.1: Classification of Plasticizers (2)

Compound Class	Issues
Phthalates	Good compatibility, high gelling capacity, low volatility
Cyclohexanoic diesters	Good substitutes for DOP (15)
Phosphates	Flame retardant, not suitable for food applications
Adipates	Low viscosity, high gelling capacity, relatively volatile and extractable, superior low temperature flexibility
Azelates	Good low temperature flexibility, less water sensitive than adipates
Sebacates	Good low temperature performance
Polymers	Very low volatility, highly resistant to extraction and migration
Trimellitates	Low volatility, good water resistance, high temperature stability,
Citrates	Good solvating power for PVC and cellulose acetate, high efficiency, non-toxic
Chlorinated hydrocarbons	Flame retardant, limited compatibility, odorous

Table 2.2: Phthalate Based Plasticizers

Compound	Examples of Uses
Diethyl phthalate	Cellulose nitrate, printing transfer films for curved-surface printing (16), conductive composite (17), orthodontic adhesives (18)
Di-*n*-butyl phthalate	Poly(vinyl acetate) (19), porosity enhancing in lithium cells (20), medical nail lacquers (21)
Di-*i*-butyl phthalate	Cellulose nitrate
Butyl benzyl phthalate	Poly(urethane) (PU) foams (22), food conveyor belts, inkjet printing media (23)
Di-*n*-hexyl phthalate	Automotive applications
Di-(2-ethylhexyl) phthalate	General Purpose, PVC
Di-*n*-octyl phthalate	
Diisooctyl phthalate	General purpose, means often Di-(2-ethylhexyl) phthalate
Diisononyl phthalate	Shoes, toys, high temperature applications (24), plastic cork stoppers (25)
Diisodecyl phthalate	Cable insulation
Diisotridecyl phthalate	Very low vapor pressure

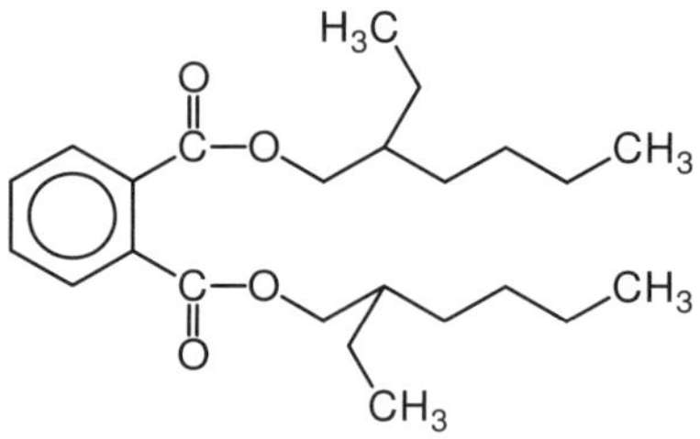

Figure 2.3: Di-(2-ethylhexyl) phthalate

they are more volatile than phthalates with long alkyl chains. The structure of DOP is shown in Figure 2.3.

Their effectiveness in plasticizing is reduced by chain branching. This effect is a stronger the nearer the branches are positioned to the carboxyl group. Further, the effect is more pronounced for shorter main chains. The branched structure accounts for a comparative increase in viscosity. For this reason, viscosity and effectiveness are closely related.

Terephthalate esters, oligoesters of o-phthalic acids, and in general, solid phthalate esters are rarely used because of their high cost.

2.5.2 Cyclohexanoic Diesters

Esters of cyclohexane polycarboxylic acids can be used as plasticizers for PVC to enable products with comparable mechanical properties to be obtained using less PVC. The use of these esters also produces formulations with increased stability to UV light, improved low temperature properties, lower viscosity, and improved processability, as well as reduced smoke on burning (15).

If esters of cyclohexane polycarboxylic acids are used as plasticizers in one of adjacent layers of plasticized PVC and phthalate plasticizers particularly DOP are used as plasticizer in the other adjacent layer, the migration of the plasticizer from one layer to the other is reduced. Undesirably high levels of migration can lead to unsightly crinkling of a multi layer foil.

A series of comparative experiments using either phthalate esters or cyclohexanoic esters have been presented (15). For example,

Table 2.3: Hardness of PVC with Different Plasticizers (15)

Plasticizer	DOP	DEHCH	DINP	DINCH	DIDP	DIDCH
phr	50	50	53	54.5	55	57.5
Shore A	90.3	90.9	91.2	91	90.4	91.7
Shore D	38.2	38	38.7	38	38	36.4

DOP: Di-(2-ethylhexyl) phthalate
DEHCH: Di-2-ethylhexyl-1,2-cyclohexane diacid ester
DINP: Diisononyl phthalate
DINCH: Diisononyl-1,2-cyclohexane diacid ester
DIDP: Diisodecyl phthalate
DIDCH: Diisodecyl-1,2-cyclohexane diacid ester

measurements of the hardness are shown in Table 2.3.

We annotate that 1,2-cyclohexanedicarboxylic anhydride can be also addressed as hexahydrophthalic anhydride. The esters of cyclohexane polycarboxylic acids may be used alone or in admixture with other plasticizers when the esters of cyclohexane polycarboxylic acids may act as viscosity depressants.

Fast fusing plasticizers may also be included. The formulations are particularly useful in the production of a range of goods from semi-rigid to highly flexible materials and are particularly useful in the production of medical materials such as blood bags and tubing.

2.5.3 Phosphate Plasticizers

Phosphates have been long known as PVC plasticizers. Phosphate based plasticizers are summarized in Table 2.4.

Most common is tricresyl phosphate. Commercial tricresyl phosphate is a mixture of the ortho, meta, and para isomers.

Phosphate based plasticizers impart flame retardant properties. The flame retardant action arises because they are capable to form polyphosphoric acids by condensation reactions on heating. The polyphosphoric acids cause enhanced charring (2).

2.5.4 Aliphatic Esters

The esters of aliphatic dicarboxylic acids, such as adipates, azelates, and sebacates exhibit high plasticizing effectiveness with PVC and

Table 2.4: Phosphate Based Plasticizers (26)

Compound
Tricresyl phosphate
Trixylyl phosphate
Triphenyl phosphate
Triethylphenyl phosphate
Diphenylcresyl phosphate
Monophenyldicresyl phosphate
Dicresylmonoxylenyl phosphate
Diphenylmonoxylenyl phosphate
Monophenyldixylenyl phosphate
Tributyl phosphate
Triethyl phosphate
Trichloroethyl phosphate
Trioctyl phosphate
Tris(isopropylphenyl)phosphate

PVAc. These types provide an excellent low temperature flexibility (2).

Likewise, monocaroboxylic acid esters of poly(ol)s show good plasticizing properties. Further, epoxidized fatty acid esters are suitable as plasticizers and as stabilizers for PVC.

These compounds are capable of forming bonds with hydrogen chloride that is ejected by the decomposition of PVC.

Trimellitates, paraffinic sulfonic acid and phenyl esters, poly(ester)s, chlorinated hydrocarbons, aliphatic or aromatic monocarboxylic acid esters such as benzoates, and a variety of elastomers are common plasticizers.

2.5.5 Polymeric Plasticizers

Polymeric or oligomeric plasticizers are advantageous due to their inherent low volatility. Therefore, they have been suggested as replacement materials for traditional plasticizers.

These materials can be tailored to be highly compatible with the host polymer. Due to the high molecular weight, leaching and volatility issues are significantly improved over traditional compounds. However, polymeric plasticizers are expensive and show

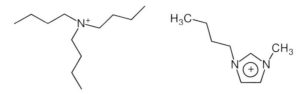

Figure 2.4: Ionic Liquids

lower plasticizing efficiency than most if the traditional plasticizers (2).

While polymeric plasticizers may cause a reduced flexibility in plastic materials, they can also be used in combination with traditional plasticizers to improve the leaching resistance.

A poly(ester) plasticizer has been described, condensed from 2-methyl-1,3-propanediol, 3-methyl-1,5-pentanediol, adipic acid, and isononanol. The poly(ester) plasticizer exhibits high plasticization efficiency and imparts excellent oil resistance to synthetic resins (27).

A triethylene glycol ester based plasticizer composition for PVC has a low heating loss, excellent adhesion, high plasticization efficiency, high elongation, high tensile strength, and high transparency (28).

2.5.6 Ionic Liquids

Ionic liquids have been investigated as plasticizers for PVC and poly(methyl methacrylate). They were found to be compatible with both the polymer systems (29). Ionic liquids are shown in Figure 2.4. Some ionic liquids, suitable as plasticizers are shown in Table 2.5.

Ionic liquids have low volatility, low melting points and a high boiling point. They are high-temperature stable, non-flammable and are compatible with a wide variety of organic and inorganic materials. A number of ionic liquids are capable of to plasticize PVC in the same manner as DOP.

16 Additives for Thermoplastics

Table 2.5: Ionic Liquids (30).

Compound
1-Butyl-3-methylimidazolium hexafluorophosphate
1-Hexyl-3-methylimidazolium dioctylsulfosuccinate
1-Hexyl-3-methylimidazolium hexafluoroborate
1-Hexyl-3-methylimidazolium hexafluorophosphate
Tetrabutyl ammonium dioctylsulfosuccinate
Tetrabutyl phosphonium dioctylsulfosuccinate
Tributyl (tetradecyl) phosphonium dodecylbenzenesulfonate
Tributyl (tetradecyl) phosphonium methanesulfonate
Trihexyl (tetradecyl) phosphonium chloride
Trihexyl (tetradecyl) phosphonium decanoate
Trihexyl (tetradecyl) phosphonium dodecylbenzenesulfonate
Trihexyl (tetradecyl) phosphonium methanesulfonate

2.6 Specific Examples of Application

Subsequently we summarize a few examples of the application and usages of plasticizers in polymeric materials. Among all kinds of additives, plasticizers are the most important class of additives for polymers. The global demand for plasticizers was 4,647 thousand metric tons in 2000 (31).

According to another study, in 2007, the global plastic additives industry grew to 12.2 million tonnes. This is justified by the rapidly growing Chinese plastics industry. Namely, China now accounts for 28% of the worldwide additives consumption (32).

2.6.1 Heat Shrinkable Films

Oriented, heat shrinkable, poly(vinylidene chloride) (PVDC) films are widely used for packaging purposes, particularly for packaging food. However, vinylidene chloride copolymers need to be plasticized so that they can be extruded and stretched into oriented films at commercial rates. The greater the proportion of plasticizer, the lower the viscosity and the easier the polymer is to extrude and orient, as well as better the abuse resistance of the final product.

Additionally, the oxygen transmission rate of the final product increases with increasing plasticizer content. However, for many purposes, it is vital that the oxygen transmission rate is low. For example, in food packaging applications, an enhanced access of oxygen would shorten the date of expire of the packed article.

Conventional plasticizers for the PVDC-methyl acrylate, are dibutyl sebacate or epoxidized soy bean oil. Glycerin together with epoxy resins is a plasticizer combination for PVDC . Another suitable plasticizer combination is a liquid epoxy resin based on epichlorohydrin and bisphenol A, and 2-ethylhexyl diphenyl phosphate (33,34). The resultant stretch oriented films show excellent oxygen permeabilities.

2.6.2 Adhesive Compositions

In poly(imide) (PI)-based adhesive compositions that are particularly useful in flexible circuit applications, plasticizers are used (35). Organic phosphates in an amount from 15–35% are added, such as triphenyl and tricresyl phosphate.

These plasticizers tend to depress the overall glass transition temperature of the PI-based adhesive, improve flame retardancy, and helps to produce both a flat and flexible coverlay coating. Other plasticizers, such as phthalate esters, aryl sulfonamides, and adipates may result in less flame resistant properties. The plasticizer is incorporated into the adhesive by dissolving it into the coating solution prior to casting and curing.

2.6.3 Interlayer Films for Safety Glasses

Poly(vinyl acetal) (PVAL) based formulations are used as interlayer films for laminated glass, as binders for ceramic forming, as binder for ink or paint and as thermally processable photographic materials (36).

Important issues are improved waterproofness and the compatibility with a plasticizer. For example, when laminate glass is exposed to high humidity for a long time, it may face problems in that water may penetrate into it through its edges and it may whiten as its compatibility with plasticizer is not good. Special formulations have been developed to overcome these drawbacks.

When the PVAL is used for interlayer films for laminated glass, a plasticizer may be added to it. A preferred plasticizer is triethylene glycol. For ceramic green sheets, dioctyl phthalate has been used as plasticizer.

The plasticizer may be added from 30–50 parts by weight. If the amount of the plasticizer added is smaller than 20 parts by weight, the interlayer films formed for laminated glass will be too tough and they could not be readily cut. However, if the added amount is larger than 100 parts by weight, the plasticizer may bleed out (36).

The backbone structure of the PVAL is crucial for the compatibility with the plasticizer. The poly(vinyl alcohol) must contain from 1–3 mol % of 1,2-glycol bond. If the 1,2-glycol bond content of PVAL is too small or too high, then the compatibility of the PVAL type becomes insufficient.

2.6.4 Electrolyte Membranes

Polymer blend membranes comprising a functional polymer based on sulfonated aryl polymers are used as polymer electrolyte membranes in fuel cells, in particular in low temperature fuel cells (37). The polymers tend to be brittle and the addition of a plasticizer which reduces the brittleness of the polymers is advantageous.

Suitable plasticizers have to be inert under the conditions prevailing in a fuel cell. Furthermore, the plasticizers have to be miscible and compatible with the functional and reinforcing polymers and be soluble in the same dipolar solvent, for example *N,N*-dimethylformamide, dimethyl sulfoxide, *N*-methyl-2-pyrrolidone (NMP) or dimethylacetamide.

According to these demands, particular preference is given to using a linear poly(vinylidene fluoride) (PVDF) as plasticizer (37). The plasticizer content is from 0.1–2% by weight.

2.6.5 Porous Electrodes

Dibutyl phthalate is used in a method of preparing an electrode for a lithium based secondary cell. In this method, $LiCoO_2$ as active material, carbon or graphite as conductive agent, PVDF as binder and dibutyl phthalate as plasticizer are mixed in an organic solvent (NMP) to prepare an electrode material composition.

The plasticizer is added to make perforations in the electrode. So, eventually the plasticizer is extracted by using an organic solvent and forms a plurality of micro-spaces in the electrode. These microspaces increase the contact area between the active material and the electrolyte (20).

2.6.6 Biodegradable Polymers

Biodegradable and other naturally degradable polymers have been attracting attention from the view point of environmental protection. One of the most important issues for the tailoring of biodegradable polymers is the rate of degradation of the product.

A number of plasticizers have been investigated for potential use in biodegradable polymers. Favorable compounds are citrate plasticizers. These are biodegradable esters.

Poly(lactic acid)s (PLA)s can be softened, for example, by adding plasticizers, blending soft polymers, or carrying out copolymerization. However, when blending soft polymers, usable soft polymers are limited to biodegradable resins such as poly(butylene succinate) from the view point of biodegradability. Such biodegradable resins have to be added in large quantities to impart sufficient flexibility, and the addition in large quantities may impair this characteristic of PLAs. Copolymerization changes the physical properties such as melting point and heat resistance, owing to the decrease in crystallinity and glass transition temperature (38).

PLA has been extensively studied in medical implants, suture, and drug delivery systems due to its biodegradability. The synthesis of several mixed alcohol esters has been described in the literature (38).

PLA has been plasticized with four commercially available citrate plasticizers: triethyl, tributyl, acetyltriethyl and acetyltributyl citrate (39).

The plasticizing effects on thermal and mechanical properties of PLA are satisfactory as the citrate esters produce flexible materials.

Some high molecular weight citrates also reduce the degradation rate of PLA. In contrast, citrate esters when used to plasticize cellulose acetate, it was found that the biodegradation rates increased dramatically with an increase in plasticizer content (40).

Additives for Thermoplastics

Table 2.6: Plasticizers for Propellants (41, 42)

Compound
Butanetriol trinitrate
Nitrocellulose
2,2-Dinitropropyl acetal
2,2-Dinitropropyl formal

2.6.7 Plasticizers for Energetic Polymers

Energetic polymers are useful in rocket propellant binder compositions, as well as in propellant compositions for air bags in the automotive industry. They are, and formed from, poly(ether)s bearing pendant azide groups crosslinked without a catalyst by a diacetylene compound (42), or triazole and tetrazole polymers, respectively.

Composite solid rocket propellants are manufactured using a variety of liquid di- and trifunctional poly(ol) prepolymers which can be crosslinked to form elastomeric PU binders which are used to form composite solid rocket fuel grains having superior mechanical properties. The PU binders are widely used in both propellants and plastic bonded explosives and were developed during the 1950's to take advantage of the long chain poly(alcohol)s which were becoming available in a wide molecular weight range. These poly(alcohol)s, when reacted with diisocyanates form stable PU polymers which could be used in large, case-bonded rocket motors. Even today, the most versatile binder systems for compounding composite propellants are derived from the reaction of hydroxyl-terminated poly(ol)s with diisocyanate to form a poly(urethane) network.

High energy nitrate esters are used to plasticize the PU propellants. Butanetriol trinitrate is such a plasticizer (41, 42). Other plasticizers for propellants are shown in Table 2.6.

Some types tend to have low values of tensile stress and modulus. This is particularly a problem with highly plasticized azido and nitrato poly(oxetane)s. These energetic polymers have sterically hindered hydroxyl groups which are slow to react with isocyanates and may not form a complete polymer network. Nitrocellulose has been added to enhance these properties, but it tends to degrade the elongation and thereby decrease toughness.

Triazole crosslinked polymers are highly effective as improved elastomeric binders for energetic compositions which may include nitrate ester plasticizers and energetic, granular fillers.

References

1. S.L. Rosen, *Fundamental principles of polymeric materials*, Wiley, New York, 1982.
2. M. Rahman and C.S. Brazel, The plasticizer market: an assessment of traditional plasticizers and research trends to meet new challenges, *Progress in Polymer Science*, 29(12):1223–1248, December 2004.
3. P.F. Bruins, ed., *Plasticizer Technology*, Reinhold Pub. Corp., New York, 1965.
4. J.K. Sears and J.R. Darby, *Technology of plasticizers*, SPE monographs, Wiley, Chichester, 1982.
5. A.S. Wilson, *Plasticizers. Principles and practice*, Vol. 585, The Institute of Materials, London, 1995.
6. B.L. Wadey, "Plasticizers," in *Encyclopedia of Physical Science and Technology*, pp. 441–456. Academic Press, New York, 2001.
7. J. Murphy, "Modifying processing characteristics: Plasticizers," in *Additives for Plastics Handbook*, pp. 169–175. Elsevier Science, Amsterdam, 2nd edition, 2001.
8. H. Zweifel, ed., *Plastics Additives Handbook*, Hanser Publishers, Munich, 5th edition, 2001.
9. H. Zweifel, R.D. Maier, and M. Schiller, eds., *Plastics Additives Handbook*, Hanser Publishers, Munich, 6th edition, 2009.
10. Standard test method for rubber properly-durometer hardness, ASTM Standard, Book of Standards, Vol. 09.01 ASTM D 2240-05, ASTM International, West Conshohocken, PA, 2005.
11. L.H. Gan, K.S. Ooi, S.H. Goh, L.M. Gan, and Y.C. Leong, Epoxidized esters of palm olein as plasticizers for poly(vinyl chloride), *European Polymer Journal*, 31(8):719–724, August 1995.
12. L.G. Krauskopf, "Monomers for polyvinyl chloride (phthalates, adipates and trimelliates)," in J. Edenbaum, ed., *Plastics Additives and Modifiers Handbook*, chapter 23, pp. 359–378. Chapman & Hall, London, 1996.
13. J.A. Tickner, T. Schettler, T. Guidotti, M. McCally, and M. Rossi, Health risks posed by use of di-2-ethylhexyl phthalate (DEHP) in PVC medical devices: A critical review, *American Journal of Industrial Medicine*, 39(1):100–111, January 2001.
14. G. Krauskopf, "Monomeric plasticizers," in E.J. Wickson, ed., *Handbook of PVC formulating*. Wiley, New York, 1993.

15. C. Gosse, T.M. Larson, P.J.P. Legrand, R.F. Caers, P.H. Daniels, A.D. Godwin, and D. Naert, Plasticised polyvinyl chloride, US Patent 7 413 813, assigned to ExxonMobil Chemical Patents Inc. (Houston, TX), August 19, 2008.
16. T. Niwa, M. Ishikawa, and K. Nakamura, Curved-surface printing method applicable to member exposed to high-temperature closed atmosphere and lamp unit having same applied thereto, US Patent 6 408 743, assigned to Cubic Co., Ltd. (Shimizu, JP) Koito Manufacturing Co., Ltd. (Tokyo, JP), June 25, 2002.
17. A. Pron, Y.F. Nicolau, M. Nechtschein, and F. Genoud, Composition for producing a conductive composite material containing a polyaniline, and resulting composite material, US Patent 6 235 220, assigned to Commissariat a l'Energie Atomique (FR), May 22, 2001.
18. K.E. Starling, Jr. and B.J. Love, Orthodontic adhesive, US Patent 6 090 867, assigned to Georgia Tech. Research Corporation (Atlanta, GA), July 18, 2000.
19. M. Mansmann, Processing of aluminum oxide fibers, US Patent 3 947 534, assigned to Bayer Aktiengesellschaft (Leverkusen, DT), March 30, 1976.
20. W. Roh, Method of preparing an electrode for lithium based secondary cell, US Patent 6 143 444, assigned to Samsung Display Devices Co., Ltd. (Kyungki-Do, KR), November 7, 2000.
21. W. Wohlrab and K. Wellner, Nail lacquer for the treatment of onychomycosis, US Patent 5 346 692, assigned to Roehm Pharma GmbH (Weiterstadt, DE), September 13, 1994.
22. J.H. Brown, A.W. Morgan, and D.S.T. Wang, Rigid polyurethane foam-forming compositions, US Patent 4 262 093, assigned to Monsanto Company (St. Louis, MO), April 14, 1981.
23. R.G. Swisher and H. Li, Inkjet printing media containing substantially water-insoluble plasticizer, US Patent 6 265 049, assigned to Hewlett-Packard Company (Palo Alto, CA), July 24, 2001.
24. A.D. Godwin, J.E. Stanat, and R.J. Saplis, Plasticizers from less branched nonyl alcohols, US Patent 6 969 736, assigned to ExxonMobil Chemical Patents Inc. (Houston, TX), November 29, 2005.
25. B. Breitscheidel, G. Olbert, K. Rossato, and U. Storzum, Single piece closure device made of pvc, US Patent 7 337 913, assigned to BASF Aktiengesellschaft (Ludwigshafen, DE), March 4, 2008.
26. K. Obuchi, M. Tokoro, T. Suzuki, H. Tanisho, and K. Otoi, Thermoplastic dicyclopentadiene-base open-ring polymers, hydrogenated derivatives thereof, and processes for the preparation of both, US Patent 6 511 756, assigned to Nippon Zeon Co., Ltd. (Tokyo, JP), January 28, 2003.

27. S. Koube, K. Iwanami, T. Arai, and T. Honda, Polyester plasticizer and chlorine-containing resin compositions, US Patent 7 348 380, assigned to Adeka Corporation (Tokyo, JP), March 25, 2008.
28. H. Kim, K. Lee, K.I. Lee, and B. Chun, Triethyleneglycol ester based plasticizer composition for polyvinyl chloride resin and method of preparing the same, US Patent 7 326 804, assigned to LG Chem, Ltd. (KR), February 5, 2008.
29. M.P. Scott, M.G. Benton, M. Rahman, and C.S. Brazel, Plasticizing effects of imidazolium salts in PMMA: High temperature stable flexible engineering materials, in R.D. Rogers and K.R. Seddon, eds., *Ionic liquids as green solvents: Progress and prospects*, Vol. 865 of *American Chemical Society Symposium Series*, pp. 468–477, Washington, DC, 2003. American Chemical Society, ACS.
30. M. Rahman and C.S. Brazel, Ionic liquids: New generation stable plasticizers for poly(vinyl chloride), *Polymer Degradation and Stability*, 91(12):3371–3382, December 2006.
31. CMR Chemical Market Resources, Inc., Global plasticizers 2000-2005 chemicals – markets, technologies & trends series, [electronic:] http://www.cmrhoutex.com/corp/multi-clients/global-plasticizers-1999-2004.htm, 2000.
32. K. Bender, Plastic 7 additives: A new multiclient research study, Townsend's Seventh Report on the Global Plastic Additives Market since 1991 7, Townsend Polymer Services & Information, Inc., Houston, TX, 2008. [electronic:] http://www.townsendsolutions.com/Portals/0/Plastic%20Additives%207%20Options%20Prospectus.pdf.
33. R.D. Moffitt, Extruding vinylidene chloride copolymer flexible packaging film, US Patent 5 147 594, assigned to W. R. Grace & Co.-Conn. (Duncan, SC), September 15, 1992.
34. S. Bekele, Vinylidene chloride composition and film made therefrom, US Patent 5 759 702, assigned to W. R. Grace & Co.-Conn. (Duncan, SC), June 2, 1998.
35. T.E. Dueber, M.W. West, B.C. Auman, and R.V. Kasowski, Polyimide based adhesive compositions useful in flexible circuit applications, and compositions and methods relating thereto, US Patent 7 220 490, assigned to E. I. du Pont de Nemours and Company (Wilmington, DE), May 22, 2007.
36. T. Kusudou, Y. Kumaki, and N. Fujiwara, Polyvinyl acetal and its use, US Patent 6 992 130, assigned to Kuraray Co., Ltd. (Kurashiki, JP), January 31, 2006.
37. W. Cui, Polymer blend membranes for use in fuel cells, US Patent 6 869 980, assigned to Celanese Ventures GmbH (DE), March 22, 2005.
38. Y. Fujita and O. Sawa, Ester compound, plasticizer for biodegradable aliphatic polyester resin, and biodegradable resin composition, US

Patent 7 166 654, assigned to Daihachi Chemical Industry Co., Ltd. (Osaka, JP), January 23, 2007.
39. L.V. Labrecque, R.A. Kumar, V. Dave, R.A. Gross, and S.P. McCarthy, Citrate esters as plasticizers for poly(lactic acid), *J. Appl. Polym. Sci.*, 66 (8):1507–1513, 1997.
40. V.P. Ghiya, V. Dave, R.A. Gross, and S.P. McCarthy, Biodegradability of cellulose acetate plasticized with citrate esters, *J. Macromol. Sci. - Pure Appl. Chem.*, 33:627–638, 1996.
41. D. Ciaramitaro, Triazole crosslinked polymers in recyclable energetic compositions and method of preparing the same, US Patent 6 872 266, assigned to The United States of America as represented by the Secretary of the Navy (Washington, DC), March 29, 2005.
42. R. Reed, Jr., Triazole cross-linked polymers, US Patent 6 103 029, assigned to The United States of America as represented by the Secretary of the Navy (Washington, DC), August 15, 2000.

3
Fillers

Fillers can be composed of inorganic and organic materials (1). Examples are given in Table 3.1 and in Table 3.2.

In addition to the inorganic fillers, organic fillers such as high molecular weight styrenes, lignins and reclaimed rubber may be used.

3.1 Surface Modification

3.1.1 Siloxanes

Carbonate based fillers as well as others can be treated to make them hydrophobic. Well-known is the treatment by means of organopolysiloxanes. Herewith H–SiO– are introduced (2). Suitable siloxanes include poly(dimethylsiloxane)s.

Breathable films can be formulated, for example in articles such as disposable nappies and similar products (3).

3.1.2 Dispersion and Coupling Additives

The adhesion of filler and polymer can be enhanced by the use of coupling additives. In many industries it is often desirable to produce polymeric compounds in which fillers are well dispersed. For example, in the rubber industry it can be desirable to produce elastomeric compounds exhibiting reduced hysteresis when compounded with other ingredients, such as reinforcing fillers, and then vulcanized (4).

The hysteresis of a rubber refers to the difference between the energy applied to deform the rubber and the energy recovered as the rubber returns to its initial, undeformed state. Interaction between

Table 3.1: Fillers and Reinforcing Agents (5)

Compound
Calcium carbonate
Silicates
Glass fibers
Glass bulbs
Asbestos
Talc
Kaolin
Mica
Barium sulfate
Metal oxides and hydroxides
Carbon black
Graphite
Wood flour
Flours or fibers of other natural products
Synthetic fibers

the elastomer molecules and the reinforcing filler is known to affect hysteresis, and it has been recognized that hysteresis and other physical properties of compounded elastomer systems can be improved by ensuring good dispersion of the reinforcing filler throughout the elastomer component. Elastomeric compounds exhibiting reduced hysteresis, when fabricated into components for constructing articles such as tires, power belts, and the like, will manifest properties of increased rebound and reduced heat build-up when subjected to mechanical stress during normal use. In pneumatic tires, lowered hysteresis properties are associated with reduced rolling resistance and heat build-up during operation of the tire. As a result, lower fuel consumption is realized in vehicles using such pneumatic tires.

Carbon black and silica are well known reinforcing fillers in rubber compounds and these, as well as other fillers, are often used as fillers for polymeric compounds in other industries, such as in thermoplastics, composites, paint, and the like. In rubber, attempts at improving filler dispersion have included high temperature mixing of carbon black-rubber mixtures in the presence of selectively-reactive promoters to promote compounding material reinforcement. Other approaches have, for example, included surface oxidation of the compounding materials or chemical modifications to the ter-

Table 3.2: Inorganic Fillers (1)

Compound
Powder fillers
Natural silicic acid Silicates Powdered talc Kaolinite Calcined clay Pyrophyllite (aluminium silicate hydroxide) Sericite (fine grained mica) Wollastonite (calcium silicate)
Carbonates
Precipitated calcium carbonate Heavy calcium carbonate Magnesium carbonate
Hydroxides
Aluminum hydroxide Magnesium hydroxide
Oxides
Zinc oxide Zinc white Magnesium oxide
Silicate fillers
Hydrated calcium silicate Hydrated aluminum silicate Hydrated silicic acid
Flaky and Fibrous fillers
Mica Basic magnesium sulfate whisker Calcium titanate whisker Aluminum borate whisker Sepiolite (magnesium silicate) Xonotlite (calcium silicate) Potassium titanate Ellestadite (metamorphosed limestone)

Table 3.3: Dispersion Agents (4)

Compound
4-(2-oxazolyl)-phenyl-*N*-methyl-nitrone (4OPMN)
4-(2-oxazolyl)-phenyl-*N*-phenyl-nitrone (4OPPN)
4-(2-thiazolyl)-phenyl-*N*-methyl-nitrone
4-(2-thiazolyl)-phenyl-*N*-phenyl-nitrone
4-(2-oxazolyl)-phenyl-N-methyl-nitrilimine
4-(2-thiazolyl)-phenyl-*N*-methyl-nitrilimine

minal end of polymers using, for example, *N,N,N',N'*-tetramethyldiamino-benzophenone (Michler's ketone), tin coupling agents, etc.

All of these approaches have focused upon increased interaction between the elastomer and carbon black compounding materials resulting in stabilized dispersion of individual carbon black aggregates and a reduction in interaggregate contacts.

Dispersion of silica filler has been of concern because polar silanol groups on the surface of the silica particles tend to self-associate, leading to re-agglomeration of the silica particles after compounding, poor silica dispersion and a high compound viscosity. The strong silica filler network results in a rigid uncured compound that is difficult to process in extrusion and forming operations. It has been recognized that improved silica dispersion can be achieved by the use during compounding, of bifunctional silica coupling agents having a moiety, such as a silyl group, reactive with the silica surface, and a moiety, such as a mercapto, amino, vinyl, epoxy or sulfur group, that binds to the elastomer. Improved silica dispersion has also been achieved by the use of elastomers that have been chemically modified at the chain end with functional groups, such as alkoxy silyl groups and the like, that chemically bind and/or interact with silica to improve dispersion.

The dispersion of fillers in polymeric compositions can be improved by the use of a polymer-filler coupling compound. These compounds are based on nitrones and nitrilimines. Examples are summarized in Table 3.3 and shown in Figure 3.1.

The polymers suitable must contain C=C unsaturations in their molecular structure and include thermoplastic polymers as well as thermosetting polymers.

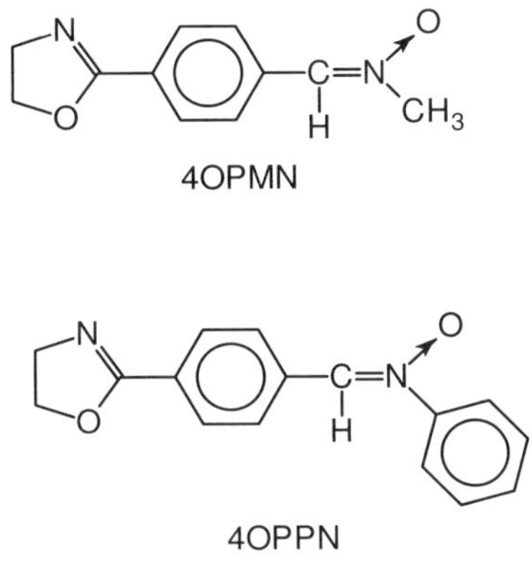

4OPMN

4OPPN

Figure 3.1: Nitrones (4)

The unsaturation can be present along the polymer backbone or be present as a pendant group, such as an ethylenic group. Suitable elastomers include both natural and synthetic rubbers. The polymer backbones of the elastomers preferably contain a significant amount of unsaturation of at least 5% of the carbon bonds.

The chemical reaction between the reactive compound and a unsaturated polymer is illustrated in Figure 3.2.

Further, the oxazolyl group and the thiazolyl group respectively, may react with silanol groups as shown in Figure 3.2. In this way a coupling of polymer and filler is achieved (4).

3.2 Special Applications

3.2.1 Flame Retardant Fillers

Flame retardant fillers for poly(imide) (PI)-based adhesive compositions consist of melamine poly(phosphate) and melamine pyrophosphate. This type of filler improves the flame retardancy of the material. The usage in PI-based adhesive compositions has been described (6).

Melamine poly(phosphate) is preferable for high temperature conditions. In contrast, melamine pyrophosphate exhibits the loss

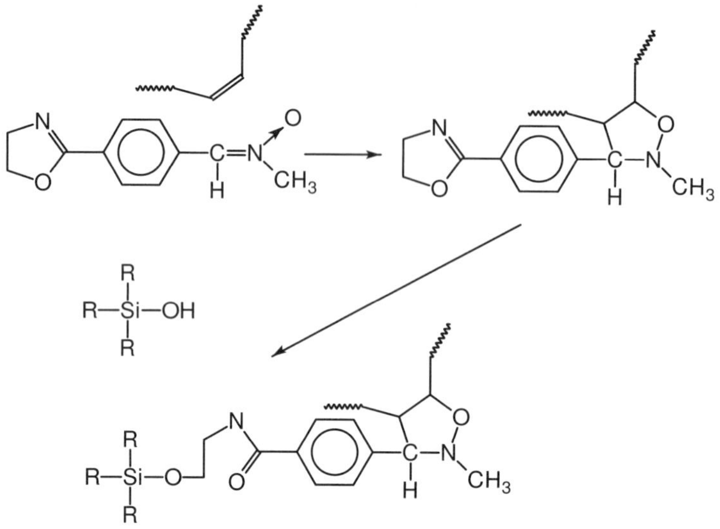

Figure 3.2: Reaction of Nitrones with Double Bonds and Bonding to Silica (4)

of water at temperatures above 300°C. This may cause blisters to form in flexible circuit laminates.

Other useful flame retardant fillers for PI-based adhesive compositions are summarized in Table 3.4.

3.2.2 Conductive Fillers

Electrically Conductive Fillers

Suitable electrically conductive fillers include carbon black, carbon fibers, vapor grown carbon fibers, carbon nanotubes, metal fillers, conductive non-metal fillers, metal coated fillers, etc (7).

Preferred carbon blacks include those having average particle sizes of less than 50 nm. Vapor grown carbon fibers should have diameters of 5–50 nm.

Carbon nanotubes may consist of a single wall, wherein the tube diameter is about 0.7–2.4 nm, or the may have multiple, concentrically-arranged walls with a tube diameter of 2–50 nm.

Conductive carbon fibers known for their use in modifying the

Table 3.4: Flame Retardant Fillers (6)

Compound Class
Ammonium poly(phosphate)
Polyphosphoric acid amide
Ammonium polyphosphoric acid amide
Melamine poly(phosphate) (Melapur200®)
Melamine poly(phosphate) acid
Melamine-modified ammonium poly(phosphate)
Melamine-modified polyphosphoric acid amide
Melamine-modified ammonium poly(phosphate)
Melamine-modified carbamyl poly(phosphate)
Carbamyl poly(phosphate)

electrostatic discharge properties of polymeric resins. Various types of conductive carbon fibers are known and classified according to their diameter, morphology, and degree of graphitization.

Usually, carbon fibers have a diameter of 3–15 μ. Carbon fibers may have graphene ribbons parallel to the fiber axis. Carbon fibers are produced commercially by pyrolysis of organic precursors such as phenolics, poly(acrylonitrile), or pitch. In general, carbon fibers are chopped, having an initial length of about 0.05–5 cm. Sized fibers are conventionally coated on at least a portion of their surfaces with a sizing composition. The sizing composition effects increased compatibility with the polymeric thermoplastic matrix material (7).

Suitable metallic fillers may be composed of aluminum, copper, magnesium, chromium, tin, nickel, silver, iron, titanium, and alloys, such as stainless steels, bronzes. In addition, metal fillers may comprise intermetallic chemical compounds, such as titanium diboride and carbides of the above metals. Further, conductive non-metal fillers may comprise tin oxide, indium tin oxide, etc (7).

Non-conductive, non-metallic fillers that have been coated over a substantial portion of their surface with a coherent layer of a solid conductive metal may also be used in conductive compositions. These types of fillers are commonly referred to as coated substrates (8).

For certain applications it may be useful to melt mix an electrically conductive fillers with e.g., poly(amide) to form a conductive masterbatch and add the conductive masterbatch to the remaining components, usually downstream of the extruder feedthroat (9).

Lithium Polymer Batteries. In lithium polymer batteries, the polymer electrolyte must satisfy requirements such as (10):

- Excellent ionic conductivity,
- Mechanical properties, and
- Interfacial stability between it and electrodes.

In particular, in lithium metal polymer batteries, dendritic growth of lithium on a lithium anode, formation of dead lithium, interfacial phenomenon between the lithium anode and the polymer electrolyte, etc., adversely affects the stability and cycle characteristics of the batteries. In view of these problems, various polymer electrolytes have been developed.

In order to manufacture a composite polymer electrolyte for a lithium secondary battery, a copolymer of vinylidene fluoride and hexafluoropropylene is casted with an conducting inorganic filler.

The conducting filler is synthesized from a hydrophobic surface treated silica that is treated with chlorosulfonic acid, $Cl-SO_3H$ in 1,2-dichloroethane. The filler is then neutralized with a lithium hydroxide, LiOH, solution. Thus eventually a lithium cationic single-ion conducting inorganic filler is obtained (10).

Thermally Conductive Fillers

Suitable thermally conductive fillers include graphite, aluminum nitride, silicon carbide, boron nitride, alumina (11).

3.2.3 Solder Precoated Fillers

Semiconductors are kept within their operating temperature limits by transferring junction generated waste heat to the ambient environment, usually the surrounding room air. This is best accomplished by attaching a heat sink to the semiconductor package surface thus increasing the heat transfer between the hot case and the cooling air. Once the correct heat sink has been selected, it must be carefully joined to the semiconductor package to ensure efficient heat transfer through this newly formed thermal interface. Thermal resistance is minimized by making the joint as thin as possible, increasing joint thermal conductivity by eliminating interstitial air and making certain that both surfaces are in intimate contact (12).

Several types of thermally conductive materials can be used to eliminate air gaps from a thermal interface, including greases, reactive compounds, elastomers and pressure sensitive adhesive films. All of these thermal interface materials (TIMs) are designed to conform to surface irregularities, thereby eliminating air voids and improving heat flow through the thermal interface.

A thermal interface material is typically made from a polymer matrix and a highly thermally conductive filler. Thermal interface materials find three application areas in a CPU package (12):

1. To bring a bare die package into contact with heat sink device,
2. To bring the die into good thermal contact with an integrated heat spreader, and
3. To bring the integrated heat spreader into contact with the electronic devices.

The fusible filler is selected from an indium alloy (12).

3.2.4 Nano Clays

Nano clays are used as fillers in thermoplastics and thermosets. The nano clay may be combined with another chemical ingredient, such as a crosslinking agent, to thereby provide a unique and overall synergistic effect on mechanical property performance (13).

A number of techniques have been described for dispersing the inorganic layered material into a polymer matrix. It has been suggested to disperse individual layers, e.g., platelets, of the layered inorganic material, throughout the polymer. However, without some additional treatment, the polymer will not infiltrate into the space between the layers of the additive sufficiently and the layers of the layered inorganic material will not be sufficiently uniformly dispersed in the polymer.

3.2.5 Mixed Matrix Membranes

In polymeric mixed matrix membrane that are used in gas separation applications microporous fillers are used. Soluble polymers of intrinsic microporosity are incorporated as microporous fillers.

The polymeric fillers exhibit a rigid rod-like, randomly contorted structure which allows them to exhibit intrinsic microporosity. These polymeric fillers of intrinsic microporosity exhibit behavior analogous to that of conventional microporous materials including large and accessible surface areas, interconnected micropores of less than 2 nm in size, as well as high chemical and thermal stability, but, in addition, possess properties of conventional polymers including good solubility and easy processability. The fillers are made from poly(ether)s that have a favorable interaction between carbon dioxide and the ethers within the chain.

Further, these polymeric fillers are reducing the hydrocarbon fouling problem of PI membranes. Gas separation experiments show a dramatically enhanced gas separation performance for CO_2 removal from natural gas.

The mixed matrix membranes can also be used in the separation of the following pairs of gases (14):

- Hydrogen–methane,
- Carbon dioxide–nitrogen,
- Methane–nitrogen, and
- Propene–propane.

References

1. T. Tanizaki, T. Nakahara, and M. Kubo, Poly (4-methyl-1-pentene) resin laminates and uses thereof, US Patent 6 265 083, assigned to Mitsui Chemicals, Inc. (Tokyo, JP), July 24, 2001.
2. R.M. Braun, P. Panek, H. Bornefeld, D. Rade, and W. Ritter, Hydrophobic pigments and fillers for incorporation into plastics., EP Patent 0 257 423, assigned to Bayer AG, March 02, 1988.
3. H.U. Hoppler, E. Ochsner, and D. Frey, Method for treating a mineral filler with a polydialkylsiloxane and a fatty acid, resulting hydrophobic fillers and uses thereof in polymers for breathable films, US Patent 7 312 258, assigned to Omya Development AG (Oftringen, CH), December 25, 2007.

4. Y. Fukushima, R.W. Koch, W.L. Hergenrother, and S. Araki, Polymer-filler coupling additives, US Patent 7 186 845, assigned to Bridgestone Corporation (Tokyo, JP), March 6, 2007.
5. P. Piccinelli, M. Vitali, A. Landuzzi, G. Da Roit, P. Carrozza, M. Grob, and N. Lelli, Permanent surface modifiers, US Patent 7 408 077, assigned to Ciba Specialty Chemicals Corp. (Tarrytown, NY), August 5, 2008.
6. T.E. Dueber, M.W. West, B.C. Auman, and R.V. Kasowski, Polyimide based adhesive compositions useful in flexible circuit applications, and compositions and methods relating thereto, US Patent 7 220 490, assigned to E. I. du Pont de Nemours and Company (Wilmington, DE), May 22, 2007.
7. J.H.P. Bastiaens, G.J.C. Doggen, and J.G.M. van Gisbergen, Conductive polyphenylene ether-polyamide composition, method of manufacture thereof, and article derived therefrom, US Patent 7 022 776, assigned to General Electric (Pittsfield, MA), April 4, 2006.
8. N.C. Patel, Methods of forming conductive thermoplastic polyetherimide polyester compositions and articles formed thereby, US Patent 6 734 262, assigned to General Electric Company (Pittsfield, MA), May 11, 2004.
9. M.D. Elkovitch, J.R. Fishburn, and S.P. Ting, Poly (arylene ether)/polyamide composition, US Patent 7 182 886, assigned to General Electric Company (Schenectady, NY), February 27, 2007.
10. Y.G. Lee, K.M. Kim, K.S. Ryu, and S.H. Chang, Lithium cationic single-ion conducting filler-containing composite polymer electrolyte for lithium secondary battery and method of manufacturing the same, US Patent 7 399 556, assigned to Electronics and Telecommunications Research Institute (Daejeon, KR), July 15, 2008.
11. S.M. Dershem, D.B. Patterson, and J.A. Osuna, Jr., Maleimide compounds in liquid form, US Patent 7 102 015, assigned to Henkel Corporation (Rocky Hill, CT), September 5, 2006.
12. P.A. Koning, F. Hua, and C.L. Deppisch, Polymer with solder precoated fillers for thermal interface materials, US Patent 7 036 573, assigned to Intel Corporation (Santa Clara, CA), May 2, 2006.
13. L.A. Acquarulo, Jr., C. O'Neil, and J. Xu, Optimizing nano-filler performance in polymers, US Patent 7 034 071, assigned to Foster Corporation (Putnam, CT), April 25, 2006.
14. C. Liu and S.T. Wilson, Mixed matrix membranes incorporating microporous polymers as fillers, US Patent 7 410 525, assigned to UOP LLC (Des Plaines, IL), August 12, 2008.

4
Colorants

4.1 Physics Behind a Color

The physics behind a color is described in monographs (1–3) as well as the sensations behind (4). The science of color is also addressed as chromatics. This includes the

1. Physics of electromagnetic radiation,
2. Origin of color in materials, and
3. Subjective perception of color.

4.1.1 Human Eye

In the human eye, there are rods and cones that act as sensors for light. In contrast to the cones, which serve as color receptors, the rods are not sensitive to color, but more sensitive to overall light. Therefore, in twilight, the perception of color is reduced. There are three types of cones that respond differently for different wavelengths. These regions are roughly red, green, and blue, respectively. The relative sensitivity of the three types of cones is schematically shown in Figure 4.1.

The subjective perception of a certain color results from the distribution of the wavelength of the incident light and the sensitivity of the eye to these particular wavelengths.

4.1.2 Tristimulus Values

The tristimulus values result from the natural sensitivity of the human eye to colors. There are three types of sensor cells in the human eye that respond to three regions of the visible light. The basic idea

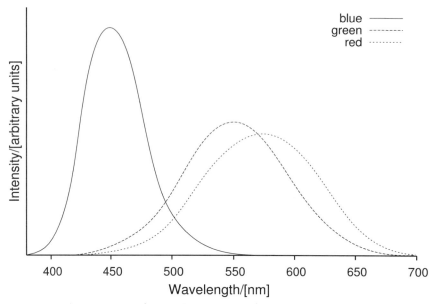

Figure 4.1: Relative Sensitivity of Human Cones (2, 5)

is that if there are three types of receptors, it should be possible to adjust each color impression by a set of only three light sources. The sources are chosen in the region of the maximum sensitivity of the three different cone types of the eye. The tristimulus values $(X, Y, Z,$ or $(R, G, B))$ for a certain color with a spectral distribution of $I(\lambda)$ are given by

$$R = \int_0^\infty I(\lambda)\bar{r}(\lambda)d\lambda$$
$$G = \int_0^\infty I(\lambda)\bar{g}(\lambda)d\lambda$$
$$B = \int_0^\infty I(\lambda)\bar{b}(\lambda)d\lambda$$

Here, $\bar{r}, \bar{g},$ and $\bar{b}$ are the response functions in order to match the sensitivity functions of the average human eye to get a certain impression of a color. Note that $\bar{r}, \bar{g},$ and $\bar{b}$ differ from the sensitivity functions, because the light sources may have – and actually have – a spectrum that is different from the human sensitivity functions. Therefore, also negative values of the response functions occur.

However, by suitable transformations the color matching functions can be kept positive. Details are beyond the scope of this text. The interested reader may consult the literature (2,5,6).

In summary, the tristimulus values are orientated according to the average sensitivity of the human eye. Light with a different spectral distribution may have the same effect on the three color receptors in the human eye. This will result in the perception of the same color.

4.1.3 Color Spaces

A lot of different color space systems have been constructed and tailored for a wide variety of applications (5,7,8).

CIE Values

The CIE values were established by the International Commission on Illumination (Commission Internationale de l'Eclairage, CIE) (5,6,9). The original CIE standard primary light sources are narrow band light sources with red at 700 nm, green at 546 nm, and blue at 436 nm. We emphasize that the latter are incidentally the bands of from mercury lamps (10).

The CIEXYZ color space is defined in that way that the X, Y, Z are normalized to one – like a mole fraction. In this way, a two dimensional color space can be opened. This is addressed as the chromaticity diagram. A schematic chromaticity diagram is shown in Figure 4.2. The boundary in the chromaticity diagram forms the monochromatic limiting case. Here the wavelengths are indicated.

The usual reference standard is the CIEXYZ color space or CIELAB color space (5,9). From this standard, other color spaces have been developed.

Other Color Spaces

The RGB color space uses additive color mixing. It is based on what kind of light needs to be emitted to produce a given color. In contrast, CMYK uses subtractive color mixing. It is used for inks or coatings and it is based on the light reflected from the substrate by which the surfaces produce a certain color.

There are special color spaces for the application in electronic monitors, used in computers and television.

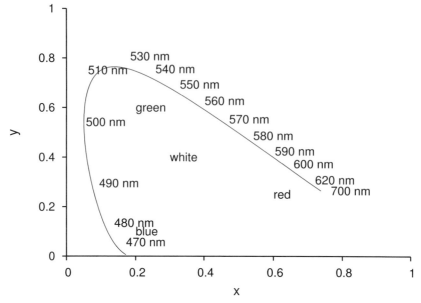

Figure 4.2: Chromaticity Diagram (schematic)

The Munsell color system specifies colors based on three color dimensions:

- Hue,
- Value,
- Chroma.

The RAL is a color matching system used in Europe and is used for varnishes and coatings. The RAL associates numbers to individual colors (11).

4.2 Color Index

The *Colour Index International* is a reference database which is maintained by the Society of Dyers and Colourists and the American Association of Textile Chemists and Colorists (12). The color index was first published in 1925. Nowadays, it is published exclusively on the internet.

The color index is a reference database for manufactured color products. It is used both by manufacturers and by consumers. The availability of a standard classification system for pigments is helpful because it resolves conflicting historic, proprietary, and generic names that have been applied to colors.

Colorants are listed in the color index according to the widely acclaimed system of color index generic names and color index constitution numbers. The numbers are prefixed with CI. For example, the compound CI 73907 has the CI generic name Pigment Red 202 and belongs to the quinacridone compounds. As another example, the compound CI 20710 has the CI generic name Pigment Yellow 93 and belongs to the azo condensation compounds. Besides generic names, trade names are available.

A detailed record of products available on the market is presented under each color index reference. Each product name is listed the manufacturer, physical form, principal usages and comments supplied by the manufacturer to guide prospective customers.

The color index numbers are grouped into ranges according to the chemical structure, as shown in Table 4.1. Sometimes an even more detailed classification is given.

4.3 Test Standards

There are numerous test standards for the determination of the color in the field of plastics available.

The yellowness and whiteness indices can be obtained from instrumentally measured color coordinates (13).

The instrumental measurement of specimens can be done resulting in color coordinates and color-difference values by using a tristimulus colorimeter. This instrument is also known as a tristimulus filter colorimeter or a color-difference meter. The test method covers both reflected and transmitted light. The method is generally suitable for any non-fluorescent, planar, object-color specimens of all gloss levels (14).

Materials pigmented with interference pigments can be characterized. by color data obtained from spectral reflectance factors at specific illumination and detection angles (15). Similarly, color information for metal flake pigmented materials, has been obtained on

Table 4.1: Color Index Numbers (16)

Moiety	Range	Category
Nitroso	10000–10299	
Nitro	10300–10999	
Monoazo	11000–19999	Azo dyes
Diazo	20000–29999	Azo dyes
Stilbene	40000–40799	
Diarylmethane	41000–41999	Diarylmethane dyes
Triarylmethane	42000–44999	Triarylmethane dyes
Xanthene	45000–45999	
Acridine	46000–46999	Acridine dyes
Quinoline	47000–47999	Quinoline dyes
Methine	48000–48999	
Thiazole	49000–49399	Thiazole dyes
Indamine	49400–49699	
Indophenol	49700–49999	Indophenol dyes
Azine	50000–50999	Azine dyes
Oxazine	51000–51999	Oxazine dyes
Thiazine	52000–52999	Thiazine dyes
Aminoketone	56000–56999	
Anthraquinone	58000–72999	Anthraquinone dyes
Indigoid	73000–73999	
Phthalocyanine	74000–74999	Phthalocyanines
Inorganic pigments	77000–77999	Inorganic pigments

paint and coatings, and the same principles should apply to plastics containing metallic flakes (17).

Standards are available for highly specialized applications. For example, there is an ASTM standard that can be used in conjunction with various thermoplastic pavement marking specifications (18).

In addition there are adequate standards in Europe dealing with color measurement (19–22).

4.4 Pigments

Pigments may be of organic and inorganic nature. Pigments are used likewise as colorants for plastics, but also are used in toner compositions. Some pigments are summarized in Table 4.2.

Some commercial names for pigments used for toners are shown in Table 4.3.

4.5 Organic Colorants

Besides of inorganic pigments, organic colorants are also useful. Polymer soluble colorants are made from substances from different organic chemical classes. The most important classes are shown in Table 4.4.

Azo compounds exhibit a limited heat resistance. Anthraquinone find use where high processing temperatures are needed.

In contrast to pigments, polymer soluble colorants are molecularly dispersed in the molten thermoplastic. For this reason, basically a high transparency and a high color strength of the final article can be achieved.

A drawback is the potential diffusion and migration in the polymer matrix. This property may lead to the permeation into adjacent materials that may be in contact with the colored polymer, including food.

Thus, polymer soluble colorants find use preferably in glassy plastic materials. Namely, glassy materials restrict the migration of the colorants.

Suitable polymers for polymer soluble colorants are styrenics, poly(ethylene terephthalate), poly(carbonate) and poly(methyl

Table 4.2: Pigments (23)

Color – Compounds
Black
Channel black Furnace black Lamp black
Magenta
2,9-Dimethyl-substituted quinacridone and anthraquinone CI Solvent Red 19
Cyan
Copper tetra-4-(octadecyl sulfonamido) phthalocyanine X-copper phthalocyanine pigment Anthradanthrene Blue Special Blue X-2137
Yellow
Diarylide yellow 3,3-dichlorobenzidine acetoacetanilides Nitrophenyl amine sulfonamide Foron Yellow SE/GLN Cl Dispersed Yellow 33 2,5-Dimethoxy-4-sulfonanilide phenylazo-4'-chloro-2,5-dimethoxy acetoacetanilide Permanent Yellow FGL

Table 4.3: Pigments for Toners (23)

Name	Producer	CI Name
Normandy Magenta RD-2400	Paul Uhlich	
Paliogen Violet 5100	BASF	Pigment Violet 19
Paliogen Violet 5890	BASF	
Permanent Violet VT2645	Paul Uhlich	
Heliogen Green L8730	BASF	Pigment Green 7
Argyle Green XP-1 11-S	Paul Uhlich	
Brilliant Green Toner GR 0991	Paul Uhlich	Pigment Green 1
Heliogen Blue L6900		Pigment Blue 15:1
L7020	BASF	Pigment Blue 15:3
Heliogen Blue D6840		Pigment Blue 15
D7080	BASF	Pigment Blue 15:3
Sudan Blue OS	BASF	
PV Fast Blue B2GO1	American Hoechst	Pigment Blue 15:3
Irgalite Blue BCA	Ciba-Geigy	
Paliogen Blue 6470	BASF	Pigment Blue 60
Sudan II	Matheson, Coleman, Bell	Solvent Red 7
Sudan III	Matheson, Coleman, Bell	Solvent Red 23
Sudan IV	Matheson, Coleman, Bell	Solvent Red 24
Sudan Orange G	Aldrich	
Sudan Orange 220	BASF	
Paliogen Orange 3040	BASF	Pigment Orange 52
Ortho Orange OR 2673	Paul Uhlich	Pigment Orange 2
Paliogen Yellow 152, 1560	BASF	Pigment Yellow 108
Lithol Fast Yellow 0991K	BASF	
Paliotol Yellow 1840	BASF	
Novoperm Yellow FG1	Hoechst	
Permanent Yellow YE 0305	Paul Uhlich	

Table 4.3 (cont): Pigments for Toners (23)

Name	Producer	CI Name
Lumogen Yellow D0790	BASF	
Sico-Gelb L1250	BASF	
Sico-Yellow D1357	BASF	Pigment Yellow 13
Hostaperm Pink E	American Hoechst	
Fanal Pink D4830	BASF	Pigment Red 51
Cinquasia Magenta	DuPont	
Lithol Scarlet D3700	BASF	Pigment Red 48:1
Toluidine Red	Aldrich	Pigment Red 3
Scarlet for Thermoplast NSD PS PA	Ugine Kuhlmann of Canada	
E. D. Toluidine Red	Aldrich	
Lithol Rubine Toner	Paul Uhlich	Pigment Red 57:1
Lithol Scarlet 4460	BASF	Pigment Red 48:2
Bon Red C	Dominion Color Comp.	
Royal Brilliant Red RD-8192	Paul Uhlich	
Oracet Pink RF	Ciba-Geigy	
Paliogen Red 3871 K	BASF	Pigment Red 123
Paliogen Red 3340	BASF	Pigment Red 226

Table 4.4: Classes of Organic Colorants (24)

Compound Class
Azo
Anthraquinone
Azaporphin
Thioindigo series
Quinacridone
Dioxazine
Naphthalenetetracarboxylic acids
Perylenetetracarboxylic acids
Phthalocyanine

methacrylate). In addition, un-plasticized poly(vinyl chloride) is a candidate. Polymer soluble colorants are often used in combination with inorganic pigments. In this way, a wide range of opaque colors can be established.

Organic colorants and the preparation of formulations have been presented in the literature (24).

There are uncountable colorants for a variety of applications, even beyond plastics in the lists on the home pages of both manufacturers (25) and artists (16).

References

1. S.K. Shevell, ed., *The Science of Color*, Elsevier Science B.V., Amsterdam, 2nd edition, 2003.
2. R.W.G. Hunt, *Measuring Colour*, Ellis Horwood Series in Applied Science and Industrial Technology, Ellis Horwood, New York, 2nd (reprint) edition, 1995.
3. K. Jack, "Color spaces," in *Video Demystified*, chapter 3, pp. 15–36. Newnes, Burlington, 5th edition, 2007.
4. F. Tebbe and C. Fath, *Color Spaces*, Jovis Verlag GmbH, Berlin, 2009.
5. G. Hoffmann, CIE color space, [electronic:] http://www.fho-emden.de/~hoffmann/ciexyz29082000.pdf, [electronic:] http://en.wikipedia.org/wiki/CIE_1931_color_space, 2006.
6. J. Schanda, ed., *Colorimetry: Understanding the CIE system*, Wiley-Interscience, New York, 2007.
7. R. Lenz, "Spectral color spaces: Their structure and transformations," in P.W. Hawkes, ed., *Advances in Imaging and Electron Physics*, Vol. 138, pp. 1–67. Elsevier, Amsterdam, 2005.
8. W. Cheetham and J. Graf, "Developing expertise: Color matching at General Electric Plastics," in I. Watson, ed., *Applying Knowledge Management*, pp. 87–120. Morgan Kaufmann, San Francisco, 2003.
9. G. Hoffmann, CIELab color space, [electronic:] http://www.fho-emden.de/~hoffmann/cielab03022003.pdf, 2008.
10. W.A. Thornton, High-pressure mercury-vapor discharge lamp having a light output with incandescent characteristics, US Patent 4 065 688, assigned to Westinghouse Electric Corporation (Pittsburgh, PA), December 27, 1977.
11. De Buijzer, RAL color space, [electronic:] http://www.unitedcomposites.net/jointpages/PDFfiles/RALcolours.pdf, 2004.

12. Society of Dyers and Colourists and American Association of Textile Chemists and Colorists, Color index international, [electronic:] http://www.colour-index.org/, 2009.
13. Standard practice for calculating yellowness and whiteness indices from instrumentally measured color coordinates, ASTM Standard, Book of Standards, Vol. 06.01 ASTM E313-05, ASTM International, West Conshohocken, PA, 2005.
14. Test method for color and color-difference measurement by tristimulus (filter) colorimetry, ASTM Standard, Book of Standards, Vol. 06.01 ASTM E1347-06, ASTM International, West Conshohocken, PA, 2009.
15. Standard practice for multiangle color measurement of interference pigments, ASTM Standard, Book of Standards, Vol. 06.01 ASTM E2539-08, ASTM International, West Conshohocken, PA, 2009.
16. D. Myers, Colour index numbers, [electronic:] http://www.artiscreation.com/Colour_Index_International_ci.htm, 2007.
17. Standard practice for multiangle color measurement of metal flake pigmented materials, ASTM Standard, Book of Standards, Vol. 06.01 ASTM E2194-03, ASTM International, West Conshohocken, PA, 2009.
18. Standard test method for evaluation of color for thermoplastic traffic marking materials, ASTM Standard, Book of Standards, Vol. 06.02 ASTM D4960-08, ASTM International, West Conshohocken, PA, 2008.
19. Colorimetric evaluation of colour coordinates and colour differences according to the approximately uniform CIELAB colour space, DIN Standard DIN 6174, DIN, Berlin, 2007.
20. Colorimetry - part 1: Basic terms of colorimetry, DIN Standard DIN 5033-1, DIN, Berlin, 2009.
21. Colorimetry: Standard colorimetric systems, DIN Standard DIN 5033-2, DIN, Berlin, 1992.
22. Colorimetry: Colorimetric measures, DIN Standard DIN 5033-3, DIN, Berlin, 1992.
23. N.A. Listigovers, F.M. Pontes, M.P. Breton, and G.K. Hamer, Ink compositions, US Patent 5 760 124, assigned to Xerox Corporation (Stamford, CT), June 2, 1998.
24. M. Kressner, J. Kolbe, F. Bremer, G. Pape, K. Wolf, and F. Kummeler, Colorant preparations, a process for their production and their use for coloring plastics, US Patent 4 332 587, assigned to Bayer Aktiengesellschaft (Leverkusen, DE), June 1, 1982.
25. Ciba, Colorants for plastics, [electronic:] http://www.ciba.com/index/ind-index/ind-automotive/products-9/ind-pla-eae-pro-colorants-3.htm, 2009.

5
Optical Brighteners

Optical brighteners are also addressed as whitening agents or fluorescent whitening agents. Optical brighteners have a wide filed of application beyond their use in polymeric formulations, including detergents, paper coatings, etc (1).

Many types of polymers absorb light in the blue spectral range. This causes a somewhat yellow appearance. However, this drawback can be repaired by the following precautions:

1. Bleaching
2. Compensation of the deficits in blue color
3. Compensation of the spectral range of reflection.

Moreover, polymers become discolored during the course of aging and service.

The traditional method to reduce the initial discoloration of the materials is to bleach them. However, this procedure carries the danger of modifying or damaging the fabric or other desired properties.

It has been known for a long time that the addition of a blue pigment, the so called *blueing*, can reduce these deficiencies. The success of this procedure arises from the fact that by the addition of a blue pigment, the reflected portion of blue light becomes more prominent. The increases of intensity in the blue spectrum admixes in the total spectrum with the reddish color and thus a more bright color is obtained. This color is the more appealing to the viewer.

5.1 Basic Principles

Optical brighteners are also referred to as fluorescent whitening agent or fluorescent brightening agent.

They provide an optical compensation for the yellow color. With optical brighteners yellowing is replaced by light emitted from optical brighteners present in the area commensurate in scope with yellow color. The violet to blue light supplied by the optical brighteners combines with other light reflected from the location to provide a substantially complete or enhanced bright white appearance. This additional light is produced by the brightener through fluorescents. Optical brighteners absorb light in the ultraviolet range 275–400 nm and emit light in the ultraviolet blue spectrum 400–500 nm (1).

The physical process of whitening of fluorescent additives follow a different mechanism. The incident UV light is absorbed by the additive and transformed into a blue light which is subsequently emitted.

Several compounds are known that can achieve brightening. The selection of a particular substance should follow the following criteria:

- Maximum whitening effect
- Hue
- Compatibility of the additive with the matrix resin
- Light fastness.

The whitening effect depends on the nature of the polymeric substrate.

The selection is also affected by its hue. In general, a neutral to blue hue is preferred. However, in Asia, sometimes a reddish hue is preferred.

Fluorescent compounds belonging to the optical brightener family are typically aromatic or aromatic heterocyclic materials often containing a condensed ring system. An important feature of these compounds is the presence of an uninterrupted chain of conjugated double bonds associated with an aromatic ring. The number of such conjugated double bonds is dependent on substituents as well as the planarity of the fluorescent part of the molecule. Most brightener compounds are derivatives of stilbene or 4,4′-diamino stilbene,

Table 5.1: Polymers that may Need Brighteners

Polymer
Poly(vinyl chloride)
Poly(styrene) and copolymers
Poly(carbonate)
Poly(urethane)
Poly(ethylene)
Poly(propylene)
Poly(methyl methacrylate)
Poly(ethylene terephthalate) fibers
Poly(amide) fibers

biphenyl, five membered heterocycles (triazoles, oxazoles, imidazoles, etc.) or six membered heterocycles (cumarins, naphthalamides, triazines, etc.) (1). Optical brighteners are used several types of polymers as summarized in Table 5.1.

5.2 Measurement

Of course, the measurement of the optical properties is important beyond the scope of plastics. There are several methods of measuring the whiteness, brightness, hue, or in general the color of plastics materials (2–5).

The method relies on the instrumental measurement of the degree of yellowness under daylight illumination of close to colorless or close to white translucent or opaque plastics. The measurement is made on pellets and is based on the tristimulus values that are obtained with a colorimeter. Methods are provided to determine the color measurements, such as yellowness index, CIE X, Y, Z, and Hunter L, a, b, or CIE L*, a*, b*. A series of specimens that is compared should have similar gloss, texture, etc.

The yellowness index is a number calculated from spectrophotometric data that describes the deviation in color of a test sample from clear or white toward yellow. The CIE values were established by the International Commission on Illumination (CIE) (6–8).

For routine evaluations, the whiteness and the tint of white samples can be calculated by empirical formulas based on colorimetric measurements colorimetry (9–11). More details concerning the physics of a color are given in Section 4.1.

Sometimes, color measurements are combined with accelerated weathering (12, 13). This procedure consists of exposing the plastic to cycles of intense light, heat and water exposure. However, no accelerated weathering test can exactly predict the performance of the material in actual outdoor exposure.

5.3 Inorganic Brighteners

Titanium pigments absorb in the near visible UV-range. Their efficiency is less in comparison to anatase pigments that are absorbing ca. 40% of the UV light in the range of wavelength of 380 nm.

5.4 Organic Optical Brighteners

Only a few classes of organic compounds are suitable as brighteners for polymers. These include:

- Bis(benzoxaziol)s
- Phenylcoumarins
- Bis-styryl biphenyls.

To be effective, the optical brightener must dissolve in the polymer. Typical addition levels of organic brighteners are in the range of 25–250 ppm.

Some organic brighteners are given in Table 5.2 and are shown in Figure 5.1.

The synthesis of an optical brightener, 2,5-bis(4'-carbomethoxystyryl)-1,3,4-oxadiazole, is shown in Figure 5.2.

Optical brighteners are used in a wide variety of polymers (14–16). There are types that have the FDA approval for a variety of food-contact applications.

5.4.1 Reactive Optical Brighteners

Since this type of optical brightener has two ester groups, it can be fixed in a polyester backbone (17). Actually, water-dispersible poly-

Table 5.2: Organic Optical Brighteners (18, p. 898)

Compound
4-Methyl-7-diethylaminocoumarin
3-Phenyl-7-(4-methyl-6-butyloxybenzoxazole)coumarin
4,4'-Bis(benzoxazol-2-yl)stilbene
2,4-Dimethoxy-6-(1'-pyrenyl)-1,3,5-triazine
1,4,-Bis(benzoxazol-2-yl)naphthalene
4,4'-Bis(2-methoxystyryl)biphenyl
2,5-Bis(5-*tert*-butyl-benzoxazol-2-yl)thiophene
Benzenesulfonic acid, 2,2'-(1,2-ethenediyl)bis(5-((4-(bis(2-hydroxyethyl)amino)-6-((4-sulfophenyl)amino)-1,3,5-triazin-2-yl)amino)-, tetrasodium salt (Blankophor®) (19)
4,4'-Bis((4-anilino-6-morpholino-1,3,5-triazin-2-yl)amino)stilbene-2,2'-disulfonate disodium salt (Tinopal®)
(2,2'-1,2-Ethenediyldi-4,1-phenylene)bisbenzoxazole (Eastobrite®)

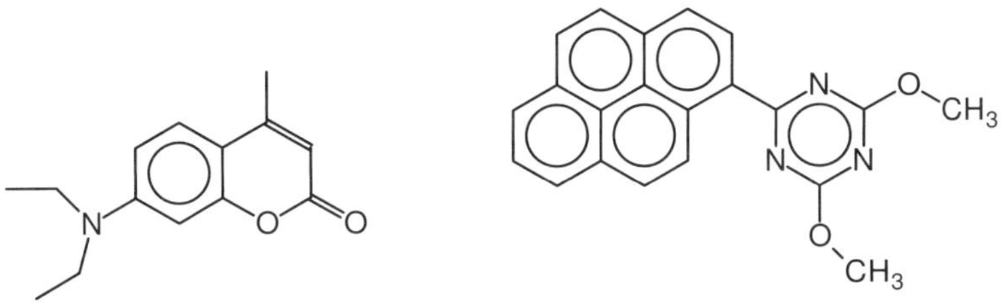

4-Methyl-7-diethylaminocoumarin   2,4-Dimethoxy-6-(1'-pyrenyl)-1,3,5-triazine

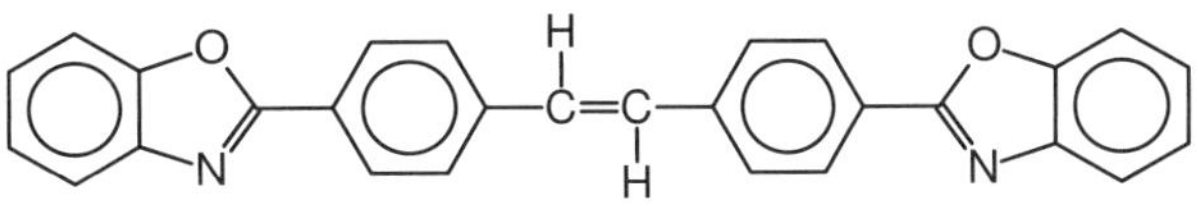

2,2'-(1,2-Ethenediyldi-4,1-phenylene)bisbenzoxazole

Figure 5.1: Organic Brighteners

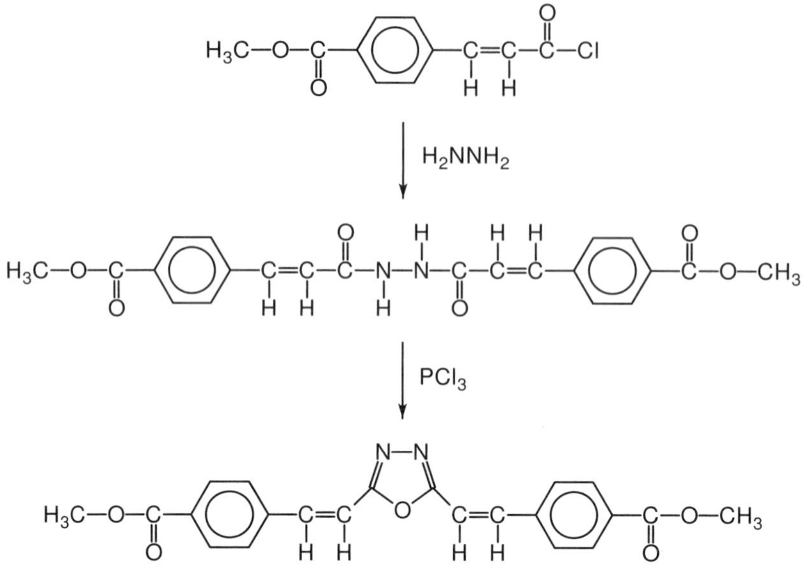

54 Additives for Thermoplastics

Figure 5.2: Synthesis of an Optical Brightener (17)

(ester)s have been described. Water-dispersible polymeric compositions are useful in the formulation of optical brightener inks, paints and film forming compositions. The water-dispersible polymeric compositions are unique in that the optical brightener compounds are not extractable, exudable, sublimable or leachable from the polymeric composition. Also, since the optical brightener compounds are bound to polymer chain by covalent bonds or incorporated into the backbone of the polymer by covalent bonds, toxicological concerns are minimized because of low potential for exposure to humans.

5.4.2 Melt Extrusion

When melt blending process or extrusion processes are applied, the optical brighteners should have a stability up to temperatures as high as 310–330°C (20). In poly(ester)s, such as poly(ethylene terephthalate) (PET) the tendency of formation of acetaldehyde can be reduced by blending with a low molecular weight poly(amide) (PA).

This is desirably for the application of PET as drinking bottles. However, the addition of PA causes yellowing of the products. Therefore, optical brighteners are added to compensate for this drawback (20).

5.4.3 Photographic Supports

A valuable class of photographic supports and elements comprises paper base material having a poly(olefin) coating containing a white pigment and an optical brightener. Such supports are particularly useful in the preparation of photographic elements such as color prints because they exhibit good brightness and excellent dimensional stability, and are highly resistant to the action of aqueous acid and alkaline photographic processing solutions. The poly(olefin) coating provides a very smooth surface which is desirable when thin layers, such as silver halide emulsion layers, are to be coated (21).

A problem that has developed in employing optical brightening agents in the poly(olefin) layer for photographic elements is that optical brightening agents have a tendency to migrate toward the surface of the poly(olefin) and exude from the surface to form a film on the surface of the poly(olefin). Such exudation not only gives rise to a non-uniform brightness of the reflection surface of the support, but also readily transfers to any other surface contacted with that surface.

This drawback can be mitigated by subjecting the free surface of the poly(olefin) layer to a corona discharge immediately after the poly(olefin) layer is coated on the paper support.

In this way the exudation of the optical brightening agent from the surface of the poly(olefin) is prevented. In addition, a second corona discharge immediately prior to coating light-sensitive emulsion layer is necessary.

It has been found that the first corona discharge treatment will prevent the exudation or blooming of the optical brightener to the surface of the poly(olefin) layer during the interim storage period between the coating of the poly(olefin) layer and the subsequent application of the various layers to convert the poly(olefin) coated paper support into a light-sensitive element. While this first corona discharge treatment will prevent the exudation of the optical brightening agent to the surface of the poly(olefin) layer, the adhesive nature of the poly(olefin) surface brought about by the corona discharge treatment will be lost over a period of time. Therefore, a second corona discharge treatment is required immediately prior to the application of subsequent layers to the poly(olefin) surface in order to restore the adhesive nature thereof (21).

References

1. E.D. Sowle and C.J. Parker, III, Presoak detergent with optical brightener, US Patent 6 099 589, assigned to Kay Chemical Company (Greensboro, NC), August 8, 2000.
2. Colouring materials in plastics: - Determination of colour stability to heat during processing of colouring materials in plastics - Part 1: General introduction, DIN Standard DIN EN 12877-1, DIN, Berlin, 1999.
3. Standard test method for color determination of plastic pellets, ASTM Standard, Book of Standards, Vol. 08.03 ASTM D6290-05, ASTM International, West Conshohocken, PA, 2005.
4. Standard practice for calculating yellowness and whiteness indices from instrumentally measured color coordinates, ASTM Standard, Book of Standards, Vol. 06.01 ASTM E313-05, ASTM International, West Conshohocken, PA, 2005.
5. Standard test method for polyurethane raw materials: Instrumental measurement of tristimulus CIELAB color and yellowness index of liquids, ASTM Standard, Book of Standards, Vol. 08.03 ASTM D7133-05, ASTM International, West Conshohocken, PA, 2005.
6. J. Schanda, ed., *Colorimetry: Understanding the CIE system*, Wiley-Interscience, New York, 2007.
7. G. Hoffmann, CIE color space, [electronic:] http://www.fho-emden.de/~hoffmann/ciexyz29082000.pdf, [electronic:] http://en.wikipedia.org/wiki/CIE_1931_color_space, 2006.
8. G. Hoffmann, CIELab color space, [electronic:] http://www.fho-emden.de/~hoffmann/cielab03022003.pdf, 2008.
9. E. Ganz, Whiteness formulas: A selection, *Appl. Opt.*, 18(7):1073–1078, April 1979.
10. E. Ganz and R. Griesser, Whiteness: Assessment of tint, *Applied Optics*, 20(8):1395–1396, 1981.
11. Colorimetry, Publication CIE 15.2, Commission Internationale de l'Eclairage, Vienna, 1986.
12. Standard practice for xenon-arc exposure of plastics intended for indoor applications, ASTM Standard, Book of Standards, Vol. 08.01 ASTM D4459-06, ASTM International, West Conshohocken, PA, 2006.

13. Standard practice for xenon-arc exposure of plastics intended for outdoor applications, ASTM Standard, Book of Standards, Vol. 08.01 ASTM D2565-99, ASTM International, West Conshohocken, PA, 2008.
14. Eastman extends optical brightener range with product for flexible PVC and polyethylene, *Additives for Polymers*, 2003(7):2–3, July 2003.
15. BASF's optical brightener, *Focus on Pigments*, 2003(9):6, September 2003.
16. Eastobrite optical brighteners. effective brighteners for plastic, Product Information AP-8C. [electronic:] http://www.eastman.com/Literature_Center/A/AP8.pdf, Eastman Chemical Company, Kingsport, TN, 2000.
17. R.H.S. Wang, J.J. Krutak, M.K. Sharma, and B.C. Jackson, Polymers containing optical brightener compounds copolymerized therein and methods of making and using therefor, US Patent 6 150 494, assigned to Eastman Chemical Company (Kingsport, TN), November 21, 2000.
18. H. Zweifel, ed., *Plastics Additives Handbook*, Hanser Publishers, Munich, 5th edition, 2001.
19. Merlin, Papierweißtöner für alle Applikationen – Blankophor®, Firmenschrift 399_08_211_Applikation_rz.indd, Kemira Germany GmbH, Innovationspark, Marie-Curie-Straße 10 D-51377 Leverkusen, Germany, May 2008.
20. J.S.N. Dalton, S.L. Stafford, and J.D. Hewa, Polyester and optical brightener blend having improved properties, US Patent 5 985 389, assigned to Eastman Chemical Company (Kingsport, TN), November 16, 1999.
21. J.F. Carroll, D.A. Griggs, and W.A. Mruk, Reduction of optical brightener migration in polyolefin coated paper bases, US Patent 5 061 610, assigned to Eastman Kodak Company (Rochester, NY), October 29, 1991.

6
Antimicrobial Additives

Various bacteria, fungi, viruses, algae and other microorganisms are known to be in the environment and to potentially adversely affect people coming in contact with them. Such microorganisms are often undesirable as a cause of illness, odors and damage to a wide variety of material and substrates. In order to combat such microorganisms, antimicrobial agents have been suggested (1).

Antimicrobial agents have mostly low molecular weight molecules that kill or suppress the growth of viruses, bacteria, or fungi. Molecular aspects on the action of antimicrobial agents have been compiled in the literature (2). An activity against viruses is of minor interest in the polymer industry. A markable exception is in the field of medical applications.

6.1 Modes of Action

Essentially two modes of activity can be differentiated:

1. Microbiostatic activity inhibits the reproduction of microorganisms. The cells are not killed, but their growth is suppressed.
2. Microbicidal activity kills the bacteria or fungi. It reduces the number of microorganisms.

Depending on the target organisms, the term microbiostatic is replaced by bacteriostatic, or fungistatic, for the first activity, The activity is named bactericidal or fungicidal for the second case. In some cases, for example, the same antimicrobial substance can act as both as microbicidal and microbiostatic agent. The mode of activity depends then on the conditions. The activity of and antimicrobial agent is influenced by the:

1. Concentration of the antimicrobial
2. pH
3. Temperature
4. Type of polymer
5. Sensitivity of the particular microorganism.

6.1.1 Types of Irritations

Main groups of microorganisms are: bacteria, gram-positive bacteria and gram-negative bacteria, fungi, and yeasts.

The type of action of various microorganisms on plastics is shown in Table 6.1. The action of various microorganisms on human health is shown in Table 6.2.

Antimicrobial additives are added to polymers in order to protect the material against the negative impact arising from microorganisms.

The first antimicrobial additives that were used in the plastic industry were arsenic, sulfur, mercury, and copper compounds. For example, the *Bordeaux mixture* is hydrated lime in copper sulfate solution. It is still used in used as a fungicide in vineyards. In the Bordeaux region of France, where this agent was invented, it is known as *Bouillie Bordelaise*. The copper ions Cu^{2+} affect the enzymes in the fungal spores as they prevent the germination process. This was the principal biocide until the 1930s. In 1934, the fungicidal activity of dithiocarbamates was discovered. This event launched the research in organic antimicrobial substances.

6.2 Plasticizers

Plasticized poly(vinyl chloride) (PVC), poly(urethane)s, poly(ethylene) (PE), and poly(ester)s are particularly sensitive to microbial attack. Flexible PVC is by far the main plastic in which biostabilizers are incorporated, followed by poly(urethane) foams, and other resins. PVC itself is resistant to microbial attack, however, plasticizers, fillers, pigments, lubricants, and other additives used in PVC are

Table 6.1: Action of Various Microorganisms on Plastics

Effect/ Microorganism	Remark/Action on
Degradation	(3, p. 648)
Fusarium spp. [a]	Poly(ester)s (4)
Aspergillus flavus	Poly(ethylene adipate), etc. (4)
Aureobasidium spp.	Poly(caprolacton) (4)
Chaetomium globosum	Poly(caprolacton) (4)
Penicillium funiculosum	Poly(3-hydroxybutyrate) (4)
Paecilomyces variotii	Poly(hydroxyalkanoate)s (4)
Odor	
Klebsiella spp.	(5)
Proteus spp.	All of the Proteus strains form during their metabolism the enzyme urease, which has the ability to rapidly break down urea into ammonia (6).
Pseudomonas spp.	
Discoloration	Microorganism/Remark (3, p. 648)
Black	Aspergillus niger, ubiquitous in black colonies
Black	Alternaria spp., grows in black or gray colonies
Black	Cladosporium herbarum, found on dead herbaceous and woody plants, textiles, rubber, etc.
Yellow	Erwinia spp., plant pathogen
Yellow	Penicillium chrysogenum, mold sometimes exudes a yellow pigment
Green	Penicillium glaucum, used in the making French blue cheese
Red	Fusarium roseum, plant pathogen
Red	Micrococcus roseus, secretes a carotenoid pigment
Red	Serratia marcescens, in wet rooms, human pathogen, slimy films with pink discoloration
Red	Rhodotorula spp., a yeast, grows in orange to red colonies
Blue	Pseudomonas aeruginosa, secretes pigments, e.g., pyocyanin

[a] spp. is a shorthand for species

Table 6.2: Action of Various Microorganisms on Human Health (3, p. 649)

Effect
Body malodor
Athlete's foot
Wound infections
Allergic reactions
Infections
Food pathogens

Table 6.3: Susceptibility of Plasticizers to Microbial Attack

Highly susceptible types
Sebacates
Epoxidized oils
Poly(ester)s
Glycolates
Moderately susceptible types
Adipates
Azelates
Pentaerythritol esters
Low susceptible types
Phthalates
Phosphates
Chlorinated hydrocarbons

possible nutrients for microorganisms. As is well known, the plasticizer in PVC can comprise more than 50%. Ester-type plasticizers are particularly good substrates for some microbial enzymes, i.e., esterases. Plasticizers can be grouped according to their susceptibility to microbial attack as given in Table 6.3.

The biodeterioration of plastics by microorganisms can occur by different mechanisms:

1. By direct degradation using the ingredients as nutrient source
2. By indirect degradation via microbial metabolites
3. By the settlement of microorganisms on the plastic without

Table 6.4: Antimicrobial Additives

Compound	Reference
Additive	(3, p. 667)
4,5-Dichloro-2-*n*-octyl-4-isothiazolin-3-one Ag-Zn-Zeolite *N*-Trichloromethylthiophthalimide Zinc pyrithione Pentachlorophenyl laurate	
Polymer grafted	(7)
1-Hexadecylpyridinium chloride Dimethyl gentian violet 5-Chloro-2-(2,4-dichlorophenoxy)phenol Biguanide compounds Poly(vinylpyrrolidone) iodine complexes ε-Poly(lysine)	 (8)
Coatings	(9)
Chlorohexidine Methylisothiazolone Terpineol Thymol	

degradation.

Plastics are often initially attacked by fungi. These attacks cause the greatest degradation. After the biodegradation proceeds, the settlement of other microorganisms is facilitated. This results in an increased secondary biodeterioration. Biodeterioration causes a reduced service time of the plastic articles.

A series of antimicrobial additives have been described that is summarized in the literature (3, p. 667). Selected compounds are summarized in Table 6.4.

Certain phosphonium salt compounds are known as a biologically active substance for their broad active spectra on bacteria, fungi, and algae similarly to various nitrogen-containing compounds (10). A few additives are shown in Figure 6.1.

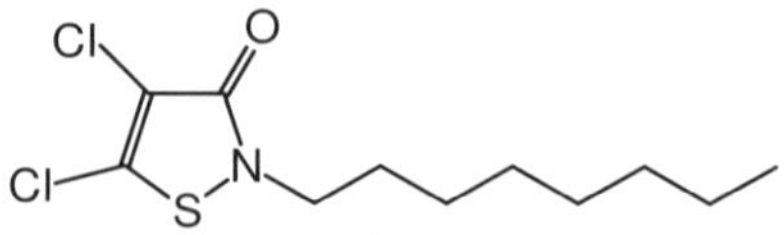

4,5-Dichloro-2-*n*-octyl-4-isothiazolin-3-one

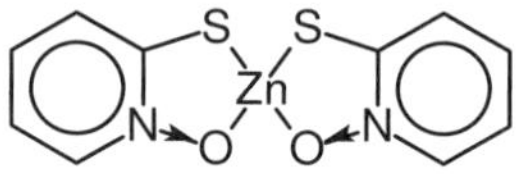

Zinc pyrithione

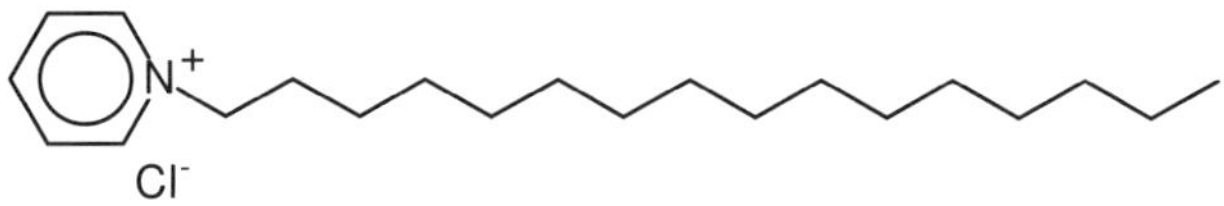

1-Hexadecylpyridinium chloride

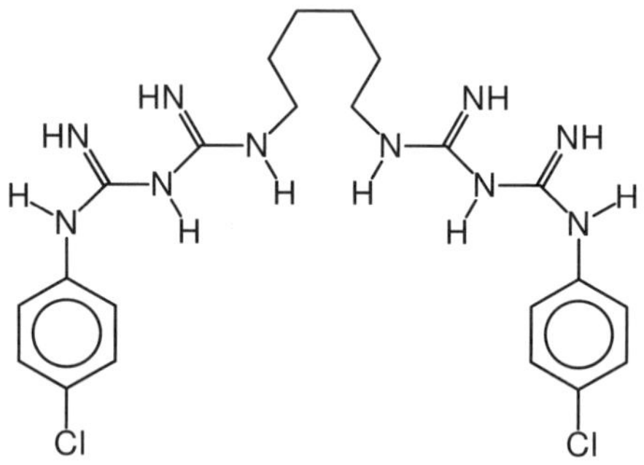

Chlorohexidine

Figure 6.1: Examples of Additives

6.3 Special Formulations

6.3.1 Contact Lenses

Contact lenses in general use, either hard or soft, are apt to be contaminated with bacteria and fungi as microorganisms easily grow between the contact lens and the cornea because of moisture, temperature and nutrients (10).

Hard contact lenses tend to mechanically damage the cornea, causing infection with microorganisms. Contact lenses for continuous wear, which have recently been increasing, are particularly dangerous. Further, bacteria, fungi, etc., may grow on the surface of hard contact lenses in a lens container, sometimes causing corneal infectious diseases.

Water-containing soft contact lenses, while comfortable to apply, are susceptible to bacteria and fungi both on the surfaces and the inside because of their own hydrophilic properties as well as their high water content, tending to cause serious infectious diseases. Moreover, they demand care in handling and involve tedious treatment for sterilization.

However, use of a large quantity of an antimicrobial agent or a potent antimicrobial agent is unfavorable to the lens and the body, especially the cornea. Therefore, research for controlling microorganisms without antimicrobials agent has been undertaken. Resins coated with an antimicrobial substance have been proposed although they are unsuitable for use as contact lens-care articles because of considerable elution of the antimicrobial substance during use.

Where antimicrobial properties are imparted to a contact lens per se, special care for safety is required; for a contact lens comes into direct contact with the cornea and the conjunctiva, and any substance eluted from the lens is carried by tears to the digestive tract. Accordingly, an antimicrobial substance to be incorporated into contact lenses and related articles is essentially required to have high activity and heat stability and to be firmly fixed so as not to be dissolved out. With these considerations in mind, phosphonium salt type polymers are deemed to be suitable.

An antimicrobial polymer can be obtained by the polymerization of an one phosphonium salt type vinyl monomer, e.g., 2-(methacrylic acid)ethyltri-*n*-octyl phosphonium chloride, tri-*n*-butylallyl phosphonium chloride, vinoxycarbonylmethyltri-*n*-butyl phosphonium chloride, and tri-*n*-octylallyl phosphonium chloride.

For example, contact lenses are prepared by the copolymerization of 2,2,3,3,4,4,4-heptafluorobutyl methacrylate, tri-*n*-octyl-3-vinylbenzyl phosphonium chloride, methyl-di(trimethylsiloxy)silylpropyl methacrylate, and ethylene glycol dimethacrylate. Isopropyl percarbonate is used as radical initiator. The resulting resin rod was cut and polished to obtain the contact lenses (10).

6.3.2 Food Packaging

Antimicrobial packaging is a form of *active packaging*. In particular, active packaging interacts with the product and the food system (11). Antimicrobial food packaging reduces, inhibits or retards the growth of microorganisms that may be present in the packed food or packaging material itself. Antimicrobial packaging includes (12):

- Addition of sachets or pads containing volatile antimicrobial agents into packages.
- Incorporation of both volatile and non-volatile antimicrobial agents directly into polymers.
- Coating or adsorbing antimicrobials onto polymer surfaces.
- Immobilization of antimicrobials to polymers by ion or covalent linkages.
- Use of polymers that are inherently antimicrobial.

The incorporation of antimicrobials into polymers has been commercially applied in drug and pesticide delivery, household goods, textiles, surgical implants and also in food-related applications.

Silver substituted zeolithes are most widely used as polymer additives for food applications (12,13). The silver is incorporated by the substitution of the sodium ions in the zeolithes. These substituted zeolites are incorporated into polymers like PE, poly(propylene), poly(amide) at levels of 1–3%.

Thermal polymer processing methods such as extrusion and injection molding may be used only with thermally stable antimicrobials. However, silver substituted zeolithes, can withstand very high temperatures, up to 800°C. Therefore, these compounds are the products of the choice to be incorporated as a thin co-extruded layer.

Heat-sensitive antimicrobials like enzymes and volatile compounds, can be formulated by solvent compounding. For example, lysozyme has been incorporated into cellulose ester films by solvent compounding in order to prevent heat denaturation of the enzyme (14,15).

In general, antimicrobial packaging materials must be in contact with the food if they are nonvolatile. The antimicrobial agents can diffuse to the surface during service. For this reason, the surface characteristics and the diffusion kinetics are important issues.

The rate of antimicrobial release from the polymer must be at a minimum rate in order to keep the surface concentration above critical concentration.

To achieve an appropriate controlled release, multilayer films have been designed (16). The matrix layer contains the active substance. The inner layer controls the rate of diffusion of the active substance. The barrier layer prevents the migration of the active agent outside of the package.

PE can be modified with maleic anhydride and on the active sites, the antimicrobial can be grafted in many cases.

6.3.3 Polymers with Inherent Antimicrobial Properties

Polymers having inherent antimicrobial or antistatic properties can be applied or used in conjunction with a wide variety of substrates such as (1):

- Textiles
- Metal
- Cellulosic materials
- Plastics.

to provide the substrate with antimicrobial properties. In addition, the polymers can also be combined with other polymers, e.g., the polymers can be used as additives)to provide such other polymers with antimicrobial properties.

Such properties are imparted by applying a coating or film formed from a cationically-charged polymer composition comprising a non-cationic ethylenically unsaturated monomer, an ethylenically unsaturated monomer capable of providing a cationic charge to the polymer composition, and a steric stabilization component incorporated into the cationically-charged polymer composition (1).

Chitosan, modified chitosans or chitosan salts can be incorporated into the compositions. Chitosan is a naturally occurring amino-functional saccharide which is known to be antimicrobial. Moreover, chitosan could also serve the dual purpose of also providing steric stabilization.

Other antimicrobial agents include metal biocides such as silver, zinc, etc. and salts. Organic compounds are chlorohexidine derivates, dodecyl diethylenediamine glycine, phenolic compounds, polymeric guanidines,

References

1. V. Krishnan, Antimicrobial and antistatic polymers and methods of using such polymers on various substrates, US Patent 7 491 753, assigned to Mallard Creek Polymers, Inc. (Chalotte, NC), February 17, 2009.
2. E.F. Gale, E. Cundliffe, P.E. Reynolds, M.H. Richmond, and M.J. Waring, *The Molecular Basis of Antibiotic Action*, John Wiley & Sons Ltd, New York, 2nd edition, 1981.
3. H. Zweifel, ed., *Plastics Additives Handbook*, Hanser Publishers, Munich, 5th edition, 2001.
4. D.Y. Kim and Y.H. Rhee, Biodegradation of microbial and synthetic polyesters by fungi, *Journal Applied Microbiology and Biotechnology*, 61 (4):300–308, May 2003.
5. P.J. Dylingowski and R.G. Hamel, "Microbial degradation of plastics," in W. Paulus, ed., *Directory of Microbicides for the Protection of Materials: A Handbook*, chapter 5.13, pp. 325–342. Springer, Dordrecht, 2005.
6. S.O. Odelhog, Germicidal absorbent body, US Patent 4 385 632, assigned to Landstingens Inkopscentral (Solna, SE), May 31, 1983.
7. W.F. McDonald, Z.H. Huang, and S.C. Wright, Antimicrobial polymer, US Patent 6 939 554, assigned to Michigan Biotechnology Institute (Lansing, MI), September 6, 2005.
8. Y. Yamamoto and J. Hiraki, Silicone-modified antimicrobial polymer, antimicrobial agent and antimicrobial resin composition, US Patent 7 470 753, assigned to Chisso Corporation (Osaka, JP), December 30, 2008.

9. R.O. Darouiche, Method of coating medical devices with a combination of antiseptics and antiseptic coating therefor, US Patent 6 162 487, assigned to Baylor College of Medicine (Houston, TX), December 19, 2000.
10. K. Hashimoto, Y. Inaba, S. Shimura, T. Mogami, T. Kojima, and Y. Ushiyama, Antimicrobial polymer, contact lens and contact lenscare articles, US Patent 5 520 910, assigned to Nippon Chemical Industrial (Tokyo, JP) Seiko Epson Corporation (Tokyo, JP), May 28, 1996.
11. A.L. Brody, E.R. Strupinsky, and L.R. Kline, *Active Packaging for Food Applications*, Technomic Publishing Co., Lancaster, PA, 2001.
12. P. Appendini and J.H. Hotchkiss, Review of antimicrobial food packaging, *Innovative Food Science & Emerging Technologies*, 3(2):113–126, June 2002.
13. P. Ackermann, M. Jaegerstaad, and T. Ohlsson, eds., *Foods and Packaging Materials – Chemical Interactions*, Royal Society of Chemistry, Letchworth, UK, 1995.
14. P. Appendini, *Immobilization of Lysozyme on Synthetic Polymers for the Application to Food Packaging*. PhD thesis, Cornell University, Ithaca, NY, 1996.
15. P. Appendini and J.H. Hotchkiss, Immobilization of lysozyme on food contact polymers as potential antimicrobial films, *Packaging Technology and Science*, 10(5):271–279, 1997.
16. J. Floros, P. Nielsen, and J. Farkas, Advances in modified atmosphere and active packaging with applications in the dairy industry, Bulletin of the International Dairy Federation 346, International Dairy Federation, Brussels, BE, 2000.

7
Flame Retardants

Polymeric materials are in general subject to burn. Therefore, for safety reasons, flame retardants are added. A measure for the flammability is the limiting oxygen index (LOI). The LOI is the percentage of oxygen in the atmosphere that allows burning under standardized conditions (1, 2). Table 7.1 gives an idea about the flammability.

A series of flame retardants, with different chemical structures exist and the mechanism of action is dependent on the nature of the particular compounds.

7.1 Mechanisms of Flame Retardants

Some important reactions in the combustion of hydrocarbons involve the following reactions:

$$\begin{aligned} H\cdot + O_2 &\rightarrow HO\cdot + O\cdot \\ O\cdot + H_2 &\rightarrow HO\cdot + H\cdot \\ HO\cdot + C\equiv O &\rightarrow CO_2 + H\cdot \end{aligned} \quad (7.1)$$

7.1.1 Flame Cooling of Halogens

Active radicals are captured by halogen X in the following manner:

$$\begin{aligned} H\cdot + HX &\rightarrow H_2 + X\cdot \\ HO\cdot + HX &\rightarrow H_2O + X\cdot \end{aligned} \quad (7.2)$$

Important for flammability is the chain branching reaction mechanism. Chain branching effects a rapid increase of the active radicals.

Table 7.1: Limiting Oxygen Index of Selected Materials (3, p.396)

Material	LOI/[% Oxygen]
Poly(formaldehyde)	15
Poly(ethylene oxide)	15
Poly(methyl methacrylate)	17
Poly(acrylonitrile)	18
Poly(ethylene)	18
Poly(propylene)	18
Poly(butadiene)	18.5
Poly(styrene)	18.5
Poly(ethylene terephthalate)	21
Poly(vinyl alcohol)	22
Poly(amide)-66	23
Wool	25
Poly(carbonate)	27
Aramid	28.5
Poly(vinyl chloride)	42
Poly(vinylidene fluoride)	44
Poly(vinylidene chloride)	60
Carbon	60
Poly(tetrafluoroethylene)	95

For this reason the rate of combustion is strongly increased, which may reach the character of an explosion. Flame cooling traps the radicals, which are notorious for accelerating the reaction rate.

Antimony trioxide Synergism

Antimony trioxide interferes with the combustion reaction as

$$\begin{aligned} Sb_2O_3 + HX &\rightarrow SbX_3 + H_2O \\ SbX_3 + H\cdot &\rightarrow SbX_2 + HX \\ SbX_2 + H\cdot &\rightarrow SbX + HX \\ SbX + H\cdot &\rightarrow Sb + HX \\ Sb + O\cdot &\rightarrow SbO\cdot \\ SbO\cdot + H\cdot &\rightarrow SbOH \\ SbOH + H\cdot &\rightarrow SbO\cdot + H\cdot \end{aligned} \quad (7.3)$$

7.2 Smoke Suppressants

Additives that impart smoke-suppressant properties to a composition tend not to be flame retardant. Conventional flame retardant and smoke-suppressant formulations include phosphorus-containing compounds such as a phosphoric acid ester, ammonium poly(phosphate) and red phosphorus, or halogen containing compounds such as tetrabromobisphenol A, decabromodiphenyl oxide and chlorinated polymers, and metal compounds such as magnesium hydroxide, aluminum hydroxide and zinc borate (4).

Halogen containing compounds exhibit good flame retardant properties and are widely used. However, compositions formulated with halogen containing compounds tend to generate undesirable levels of smoke when combusted. Other inorganic flame retardant additives include hydrated inorganic compounds, which function by absorbing heat and evolving water vapor or steam. The vapor or steam dilutes the combustible gases that are generated during a fire.

Zinc borate is considered as an environmentally friendly flame retardant and also as a smoke-suppressant. It is used as an (5, 6). Although it has been used for many polymers as flame retardants, it is a new additive for polyurethanes (5). For example, $2ZnO \times 3B_2O_3 \times 3H_2O$ has important industrial applications as smoke-suppressant (7).

Among the metallic hydroxide flame retardants, aluminium trihydrate, $Al(OH)_3$ or magnesium hydroxide $Mg(OH)_2$ are popular flame retardants and smoke suppressants (8).

Intumescent flame retardants are of interest for poly(propylene) and poly(olefin)s in general. In the case of fire, they develop very low smoke and toxic gases. Further, they reduce the notorious dripping during burning (9).

Instead, these type of flame retardants exhibit some drawbacks compared with bromine-containing flame retardants, such as low flame retardant efficiency and low thermal stability. Improved intumescent flame retardant systems have been found by formulating with zeolites and organoboron siloxanes (9).

In particular in halogen-free systems, there are major advantages with respect to smoke suppression, afterglow suppression, corrosion inhibition, by combining zinc borates with other flame retardants in certain polymers, such as ethylene vinyl acetate, poly(vinyl chloride), poly(amide) (PA).

A few studies are available that deal with the synergistic effect of zinc borate in combination with intumescent systems (10–12).

7.3 Admixed Additives

In thermoplastic resins, the flame retardants are usually admixed, but not chemically bound onto the polymeric backbone. A wide variety of flame retardants are available for thermoplastics. Halogen based flame retardants were heavily in use, but are now increasingly replaced by non-halogen containing flame retardants for environmental and health reasons.

Examples of halogenated flame retardants are shown in Table 7.2.

The synthesis of hexabromocyclododecane is shown in Figure 7.1. Hexabromocyclododecane is obtained by the trimerization of 1,3-butadiene followed by bromination. It is the most important flame retardant for foamed poly(styrene). In addition, sulfonic salt based flame retardants are shown in Table 7.3.

Organic phosphate compounds may be aromatic-based phosphates. Examples are shown in Table 7.4.

Table 7.2: Halogenated Flame Retardants (13)

Compound
Hexabromocyclododecane
Bis(2,6-Dibromophenyl)methane
1,1-Bis-(4-iodophenyl)ethane
2,6-Bis(4,6-dichloronaphthyl)propane
2,2-Bis(2,6-dichlorophenyl)pentane
Bis(4-Hydroxy-2,6-dichloro-3-methoxyphenyl)methane
2,2-Bis(3-bromo-4-hydroxyphenyl)propane
2,2'-Dichlorobiphenyl
Polybrominated 1,4-diphenoxybenzene
2,4'-Dibromobiphenyl
2,4'-Dichlorobiphenyl
Decabromodiphenyl oxide

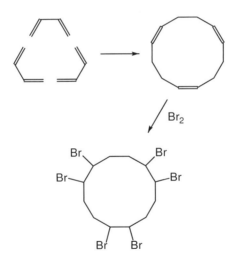

Figure 7.1: Synthesis of Hexabromocyclododecane

Table 7.3: Sulfonic Salt Flame Retardants (13)

Compound	Remark
Potassium perfluorobutane sulfonate	Rimar salt
Potassium perfluorooctane sulfonate	
Tetraethylammonium perfluorohexane sulfonate	
Potassium diphenylsulfone sulfonate	

Additives for Thermoplastics

Table 7.4: Organic Phosphate Compounds (13)

Compound
Diphenyl pentaerythritol diphosphate
Phenyl bis(dodecyl)phosphate
Phenyl bis(neopentyl)phosphate
Phenyl bis(3,5,5'-trimethylhexyl)phosphate
Ethyl diphenyl phosphate, 2-Ethylhexyl di(*p*-tolyl)phosphate
Bis(2-ethylhexyl)-*p*-tolyl phosphate
Tritolyl phosphate
Bis(2-ethylhexyl)phenyl phosphate
Tri(nonylphenyl)phosphate
Bis(dodecyl)-*p*-tolyl phosphate
Dibutyl phenyl phosphate
2-Chloroethyl diphenyl phosphate
p-Tolyl bis(2,5,5'-trimethylhexyl)phosphate
2-Ethylhexyl diphenyl phosphate
Triphenyl phosphate
Tricresyl phosphate
Isopropylated triphenyl phosphate

Table 7.5: Nitrogen–Phosphate Compounds (13)

Compound
Phosphonitrilic chloride
Phosphoric acid amides
Phosphonic acid amides
Phosphinic acid amides
Tris(aziridinyl) phosphine oxide

A phosphorus-nitrogen synergism is seen in the flame retardant action. Therefore, flame retardant compounds containing phosphorus-nitrogen bonds have been developed. Examples are shown in Table 7.5.

Inorganic salts are likewise active as flame retardants. Examples are shown in Table 7.6.

Table 7.6: Inorganic Salts (13, 14)

Compound	Remark
$CaCO_3$	Acts as diluent, lowers the combustible portion
$BaCO_3$	Inert filler
$Mg(OH)_2$	Breaks down endothermically
$Al(OH)_3$	Breaks down endothermically
KBF_4	(15)
Na_3AlF_6	(16)

Table 7.7: Bound Additives

Compound	Remark
Tetrabromobisphenol A	
Chlorendic anhydride	
Tetrabromo phthalic anhydride	

7.4 Bonded Additives

Bonded additives are bonded into the backbone or as side chains into the polymer. This type of flame retardant is prevalent in thermosetting resins. Examples are shown in Table 7.7.

Some bonded additives are shown in Figure 7.2.

7.4.1 Examples of Polymers

Poly(amides)

Combinations of organohalogen compounds and antimony compounds are well known and widely used commercially for flame retarding a great many resins, and have generally been found to be highly acceptable for use with aliphatic PAs such as nylon 6,6 and the like. However, the combination of a halogenated poly(styrene) or halogenated poly(phenylene oxide) with sodium antimonate is generally reasonably stable only at processing temperatures of up to about 300°C. Many of the commercially available flame retardants such as polybrominated poly(styrene) are described by their suppliers as being not recommended for use at temperatures above about 320°C because of the extensive decomposition that occurs at these high temperatures. Because high temperature PAs generally

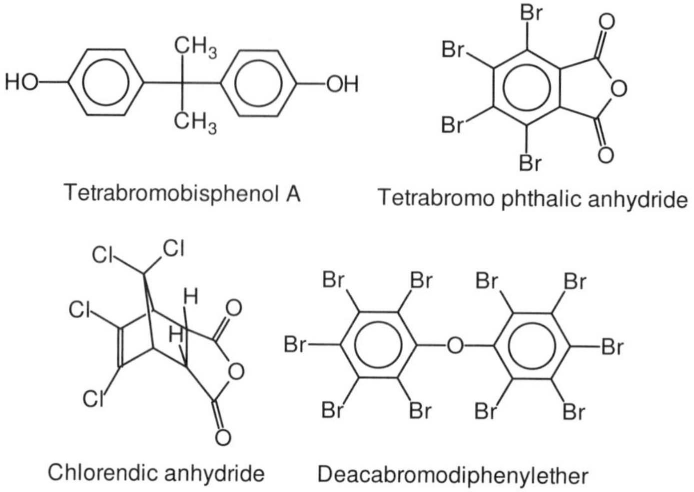

Figure 7.2: Some Bonded Additives

are melt fabricated at temperatures well in excess of 300°C, indeed as high as 350°C, particularly when filled, thermal decomposition of the flame retardant frequently occurs, which in turn inevitably causes the molded article to be discolored and detrimentally affecting the mechanical properties of the resin composition (17).

In order to improve the heat stability and thereby eliminate or minimize thermal decomposition during molding, the flame retardant high temperature PA formulations may further comprise finely divided calcium oxide. The amount of calcium oxide necessary to effect improvement in the heat stability may depend in part upon the particular PA employed, as well as upon the particular combination of antimony oxide and halogen containing flame retardant selected.

Generally, the amount is in the range of 1%, based on total weight of the composition, including such fillers, additives and fiber as may also be present. At lower levels of calcium oxide, the improvement may be found to be slight small. At high levels of calcium oxide, levels greater than about 2%, the improvement in thermal stability may be observed, but the presence of unneeded particulate materials can detrimentally affect other properties including mechanical properties, and hence these levels are not preferred.

Poly(phenylene ether) Resins

Poly(phenylene ether) (PPE) resins are high performance engineering thermoplastics having relatively high melt viscosities and softening points, i.e., in excess of 250°C. They are useful for many commercial applications requiring high temperature resistance and can be formed into films, fibers and molded articles. For example, they can be used to form a wide range of products including household appliances, automotive parts and trim (18).

PPE resins may be utilized alone or in admixture with styrene polymers over a broad spectrum of proportions. The resulting blends can be molded into many of the same articles made only from PPE resins, but with the advantage that the moldings often possess better physical or chemical properties.

Foamable compositions of PPE resins with or without styrene polymers are particularly suited as sources of lightweight structural substitutes for metals, especially in the automotive industry.

Compositions of PPE resin and styrene polymers are not normally flame retardant, and there are instances when it is desirable to impart a degree of flame retardancy to the compositions such that the molded articles are better able to resist burning or melting when exposed to elevated temperatures or placed near an open flame.

Halogenated flame retardants, e.g., brominated styrene are preferred over phosphate based flame retardants because the former exhibit less stress cracking. However, initial attempts to incorporate halogenated flame retardants into composites and then to foam such compositions, have not been successful because the compositions tend to degrade, often giving streaked articles, and generating odors despite efforts to use a variety of foaming agents. This is especially a problem when attempting to produce large sized articles by foaming the high shot weights.

However, it was discovered that a foaming agent comprising a combination of citric acid and sodium bicarbonate provides an effective means of making moldable compositions derived from PPE resins, styrene polymers and halogenated flame retardants without the undesirable by-products which lead to the decomposition of the moldable composition. Moreover, the composition is uniquely suitable to produce large size foamed shaped articles free of streaking and odor generation.

80 Additives for Thermoplastics

Table 7.8: Foamed Flame Retardant Poly(phenylene ether) Compositions (18)

Experiment	1	2
Composition	Parts by weight	
Poly(2,6-dimethyl-1,4-phenylene ether)	50	50
Rubber Modified Poly(styrene)	50	50
Polybrominated Poly(styrene)	9.5	9.5
Antimony Oxide	3.3	3.3
Butylene o-phthalate plasticizer	7.5	7.5
Styrene-ethylene butylene block copolymer impact modifier	5	5
Rubber modified poly(styrene)	1	1
Zinc sulfide stabilizer	0.15	0.15
Zinc oxide stabilizer	0.15	0.15
Titanium dioxide pigment	5	5
Citric acid–sodium bicarbonate	1	0
Properties		
Surface streaking	none	brown
Odor generation	none	substantial

A foamable composition is shown in Table 7.8.

Brominated Poly(phenylene ether)

A brominated poly(phenylene oxide) can be obtained by condensing tribromophenols (19). The reaction is shown in Figure 7.3.

The brominated poly(phenylene oxide) is used as a flame retardant. When it is added to resin, the resin moldings are excellent in

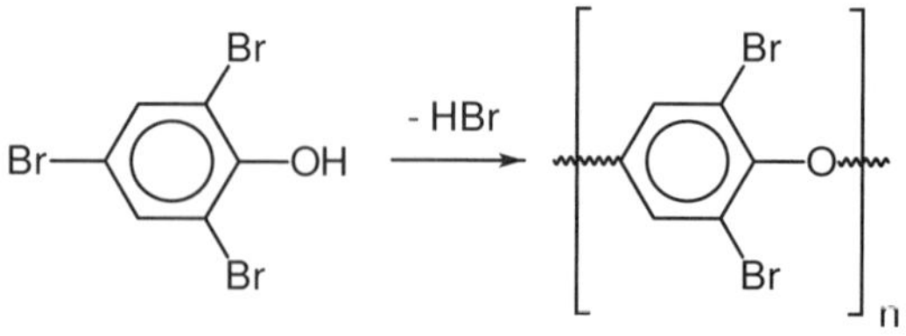

Figure 7.3: Condensation of 1,2,6-Tribromophenol

its flame retardancy, electric properties, physical properties, thermal stability and appearance (color hue), and they do not corrode molds.

Unsaturated Poly(ester)s

Unsaturated poly(ester) resins are polycondensation products made from saturated and unsaturated dicarboxylic acids or their anhydrides with diols. They are cured by free radical polymerization using initiators, e.g., peroxides and accelerators. The double bonds of the poly(ester) chain react with the double bond of the copolymerizable solvent monomer. The most important dicarboxylic acids are maleic anhydride, fumaric acid and terephthalic acid. The most frequently used diol is 1,2-propanediol, but ethylene glycol, diethylene glycol and neopentyl glycol, inter alia, are also often used (20).

The most suitable crosslinking monomer is styrene. Styrene can be mixed freely with the poly(ester) resins and is easily copolymerized. The styrene content in unsaturated poly(ester) resins is normally from 25–40%. In free-flowing unsaturated poly(ester) resins, the monomer used is usually diallyl phthalate.

Unsaturated poly(ester) resins are often converted into moldings. Moldings made from glass-fiber-reinforced unsaturated poly(ester) resins are distinguished by their good mechanical properties, low density, resistance to chemicals and excellent surface quality. This and their favorable price have allowed them increasingly to replace metallic materials in applications in rail vehicles, buildings and aeronautics. Depending on the application sector, there are different requirements with regard to mechanical, electrical and fire protection properties. Especially in the rail vehicles sector, the fire protection requirements have recently been tightened.

Unsaturated poly(ester) resins may be made flame retardant by using bromine-containing or chlorine-containing acid components or alcohol components, for example hexachloroendomethylenetetrahydrophthalic acid (HET acid), tetrabromophthalic acid or dibromoneopentyl glycol. Antimony trioxide is frequently used as synergist.

A disadvantage of bromine- or chlorine-containing poly(ester) resins is that corrosive and possibly environmentally significant gases are produced when a fire occurs and can lead to considerable damage to electronic components, for example to relays in rail vehicles. Under unfavorable conditions, polychlorinated and brominated dibenzodioxins and furans may be produced.

Unsaturated poly(ester) resins, in the form of molding compositions, may be provided with fillers such as aluminum hydroxide, which have a quenching action. At filling rates of from 150–200 parts of aluminum hydroxide per 100 parts of unsaturated poly(ester) resin, it is possible to achieve self-extinguishing and a low smoke density.

However, such formulations cannot be used for injection processes, since homogeneous distribution of the aluminum hydroxide cannot be achieved with the reinforcing materials used. For injection processes, chlorinated or brominated unsaturated poly(ester) resins are used.

The use of phosphorus compounds in unsaturated poly(ester) resins in order to establish adequate flame retardancy has already been proposed in a number of ways. The use of phosphoric esters in halogen containing unsaturated poly(ester) resins has been described (21). Melamine diphosphate has also been tested as a flame retardant for unsaturated poly(ester) resins (22). In addition, 2-methyl-2,5-dioxo-1-oxa-2-phospholane has been proposed as flame retardant (20).

Epoxide Resins

In general, a synthetic resin molded article which requires flame retardancy is blended with any of the various flame retardants such as a phosphorus-based flame retardant and a halogen-based flame retardant. These flame retardants are required to be resistant to heat, acting in the case of molding, and processing a synthetic resin or using a product after molding. At the same time, the performance of the synthetic resin should not be impaired, such as water resistivity and physical performances (23).

Phosphorus-containing compounds may suffer from such defects that impair physical characteristics of a synthetic resin, deteriorate the stability and water resistivity, etc.

This deterioration comes about because most of the compounds have each been an additive-type flame retardant. Since most of the phosphorus-containing compounds that are used as an additive-type flame retardant are poor in compatibility with a resin, about the problem that the compounds are difficult to homogeneously blend in the resin occurs. Even if homogeneous blending is possible, the problem that the flame retardant bleeds out from the molded article after molding, causes problems with the working maintenance as well as the external appearance.

A phosphorus-containing epoxy resin has been proposed, in which phosphorus-containing groups are chemically bonded to the resin matrix after molding.

Thus the flame retardant exhibits excellent water resistivity and stability, while maintaining resin characteristics to some extent. The reaction product of 9,10-dihydro-9-oxa-10-phosphaphenanthrene-10-oxide, itaconic acid in acetonitrile as solvent contains a reactive carboxylic group (23).

Poly(carbonate)

In poly(carbonate)s oligomeric and polymeric halogenated aromatic compounds, such as a co-poly(carbonate) of bisphenol A and tetra-bromobisphenol A are useful monomers (13). Halogen containing flame retardants are generally used in amounts of up to 50 phr.

References

1. Standard test method for measuring the minimum oxygen concentration to support candle-like combustion of plastics (oxygen index), ASTM Standard, Book of Standards, Vol. 08.01 ASTM D 2863-08, ASTM International, West Conshohocken, PA, 2008.
2. Kunststoffe - Bestimmung des Brennverhaltens durch den Sauerstoff-Index - Teil 2: Prüfung bei Umgebungstemperatur, DIN Standard DIN EN ISO 4589-2, DIN, Berlin, 2006.
3. G. Menges, E. Haberstroh, W. Michaeli, and E. Schmachtenberg, *Werkstoffkunde Kunststoffe*, Hanser Publishers, München, 5th edition, 2002.
4. M.J. Chanak, C.L.Y. Chase, and V.H. Batdorf, Smoke suppressant hot melt adhesive composition, US Patent 7 205 047, assigned to H.B. Fuller Licensing & Finance Inc. (St. Paul, MN), April 17, 2007.
5. B. Yildiz, M.z. Seydibeyoğlu, and F.S. Güner, Polyurethane-zinc borate composites with high oxidative stability and flame retardancy, *Polymer Degradation and Stability*, 94(7):1072–1075, July 2009.

6. S. Wallström, K. Dowling, and S. Karlsson, Development and comparison of test methods for evaluating formation of biofilms on silicones, *Polymer Degradation and Stability*, 78(2):257–262, 2002.
7. Y.H. Gao, Z.H. Liu, and X.L. Wang, Hydrothermal synthesis and thermodynamic properties of 2 $ZnO \times 3B_2O_3 \times 3H_2O$, *The Journal of Chemical Thermodynamics*, 41(6):775–778, June 2009.
8. S. Hamdani, C. Longuet, D. Perrin, J.M. Lopez-cuesta, and F. Ganachaud, Flame retardancy of silicone-based materials, *Polymer Degradation and Stability*, 94(4):465–495, April 2009.
9. Y. Li, B. Li, J. Dai, H. Jia, and S. Gao, Synergistic effects of lanthanum oxide on a novel intumescent flame retardant polypropylene system, *Polymer Degradation and Stability*, 93(1):9–16, January 2008.
10. G. Fontaine, S. Bourbigot, and S. Duquesne, Neutralized flame retardant phosphorus agent: Facile synthesis, reaction to fire in pp and synergy with zinc borate, *Polymer Degradation and Stability*, 93(1):68–76, January 2008.
11. Z. Wu, W. Shu, and Y. Hu, Synergist flame retarding effect of ultrafine zinc borate on LDPE/IFR system, *Journal of Applied Polymer Science*, 103(6):3667–3674, March 2007.
12. F. Samyn, S. Bourbigot, S. Duquesne, and R. Delobel, Effect of zinc borate on the thermal degradation of ammonium polyphosphate, *Thermochimica Acta*, 456(2):134–144, May 2007.
13. N. Agarwal, Thermoplastic polycarbonate compositions, articles made therefrom and method of manufacture, US Patent 7 498 401, assigned to SABIC Innovative Plastics IP B.V. (Bergen op Zoom, NL), March 3, 2009.
14. K. Ueno, Flame retardant and smoke suppressed polymeric composition and electric wire having sheath made from such composition, US Patent 5 059 651, assigned to Sumitomo Electric Industries, Ltd. (Osaka, JP), October 22, 1991.
15. M.A. Kasem and H.R. Richards, Flame-retardants for fabrics. function of boron-containing additives, *Ind. Eng. Chem. Prod. Res. Dev.*, 11(2):114–133, June 1972.
16. S.R. Kim, Y.J. Choi, and J.S. Song, Flame retardant polymer resin composition having improved heat distortion temperature and mechanical properties, US Patent 5 864 004, assigned to Samyang Corporation (Seoul, KR), January 26, 1999.
17. M.G. Reichmann, Flame retardant high temperature polyphthalamides having improved thermal stability, US Patent 5 773 500, assigned to Amoco Corporation (Chicago, IL), June 30, 1998.
18. E. Pressman, Modified flame retardant polyphenylene ether resins having improved foamability and molded articles made therefrom,

US Patent 4 791 145, assigned to General Electric Company (Selkirk, NY), December 13, 1988.
19. H. Onishi, Brominated polyphenylene oxide and flame retardant employing the brominated polyphenylene oxide, US Patent 6 864 343, assigned to Dai-Ichi Kogyo Seiyaku Co., Ltd. (Kyoto, JP), March 8, 2005.
20. S. Horold and H.P. Schmitz, Flame-retardant unsaturated polyester resins, US Patent 6 639 017, assigned to Clariant GmbH (Frankfurt, DE), October 28, 2003.
21. P.J. Apice, Unsaturated polyester resins prepared from phosphoric acid esters, US Patent 3 433 854, assigned to Allied Chem., March 18, 1969.
22. W. Krolikowski, W. Nowaczek, and P. Penczek, Vermindern der Brennbarkeit von ungesättigten Polyesterharzen, *Kunststoffe*, 77(9): 864–869, 1987.
23. T. Fujimura, K. Waki, K. Sato, and S. Mibae, Phosphorus-containing carboxylic acid derivatives process for preparations thereof and flame retardant, US Patent 7 186 784, assigned to NOF Corporation (Tokyo, JP), March 6, 2007.

8
Lubricants

Archeological evidence dating to before 1400 B.C. shows the use of tallow to lubricate chariot wheel axles. Leonardo da Vinci discovered the fundamental principles of lubrication and friction, but lubrication did not develop into a refined science until the late 1880's in Britain when Tower produced his studies on railroad car journal bearings in 1885. In 1886 Reynolds developed this into a theoretical basis for fluid film lubrication (1).

In the field of thermoplastic polymers, lubricants are processing aids for plastics processing. Thermoplastic resins are processed at high temperatures. However, the melt viscosity is often not sufficient to allow easy processing. Therefore, lubricants are added.

Lubricants can be active as they either are reduce the melt viscosity of the polymer or reduce the heterogeneous friction of the polymer to the static and moving walls of the processing equipment. A lubricant makes the surface of a final resin product composition smooth so as to improve its processability. Both an external lubricant and an internal lubricant can be used. The internal lubricant resides inside the polymer to reduce the viscosity of the resin itself, thereby improving flowability. The external lubricant reduces extrusion load between the polymer melt and metal surface in an extruder (2).

Apart from the usage mentioned above, polymeric compositions with lubricants are used in the field of sinter metal technology (3).

8.1 Principle of Action

The flow profile of a polymer with and without lubricant is shown schematically in Figure 8.1. There is a relationship between the

Additives for Thermoplastics

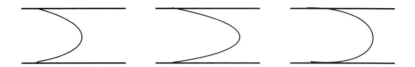

Figure 8.1: Flow Profile of a Polymer with and without Lubricant: Left without Lubricant, Center with internal Lubricant Right with external Lubricant

solubility of lubricants and their mode of action.

Basically, lubricants that are highly soluble in the polymer are addressed as internal lubricants and lubricants that are scarcely soluble are addressed as external lubricants. Further, most of the classes of lubricants are surface active substances.

The lubricating effect depends on the structure. For example, for poly(vinyl chloride) (PVC), dimethyl phthalate acts as solvent, whereas dioctyl phthalate acts as plasticizer. Eventually, distearyl phthalate acts as lubricant.

8.2 Methods of Incorporation

There are two basic methods of incorporating lubricants into the polymer, i.e., either by simple admixture, or by separate delivery into the extrusion engine. Below, I discuss the methods of incorporation in detail.

8.2.1 Conventional Method

Conventionally, the additive is admixed to the polymer and then the process of forming is started. In order to optimize the flow properties, several kinds of lubricants are admixed and used together.

8.2.2 Separate Delivery of the Lubricant

A special extrusion process has been described that involves the delivery of a lubricant separately from a polymer melt stream to each orifice of an extrusion die so that the lubricant preferably encases the polymer melt stream as it passes through the die orifice (4).

Table 8.1: Classes of Lubricants

Compound class
Alcohols
Metal soaps
Amides
Esters
Paraffin waxes
Poly(ethylene) waxes

The use of a lubricant delivered separately from the polymer melt stream in a polymeric fiber extrusion process can provide a number of potential advantages. For example, the use of separately-delivered lubricant can provide for oriented polymeric fibers in the absence of pulling, i.e., in some embodiments it may not be necessary to pull or stretch the fiber after it exits the die to obtain an oriented polymeric fiber.

If the polymeric fibers are not pulled after extrusion, they need not exhibit substantial tensile stress-carrying capability in the semi-molten state that they are in after exiting the die. Instead, the lubricated extrusion methods of the present invention can, in some instances, impart orientation to the polymeric material as it moves through the die such that the polymeric material may preferably be oriented before it exits the die.

8.3 Types of Lubricants

From the chemical perspective, lubricants are essentially waxes or fat derivatives. This implies that a carbon backbone in the range of 10–70 carbon atoms is usual. Types of lubricants are summarized in Table 8.1. Below, I discuss some issues of specific classes of lubricants.

8.3.1 Alcohols

Long chain alcohols can be produced by the Ziegler process. However, these products exhibit a high volatility. The volatility decreases when the alcohols are esterified. However, lubricants bearing alcohol groups in general tend to interact with stabilizers for PVC.

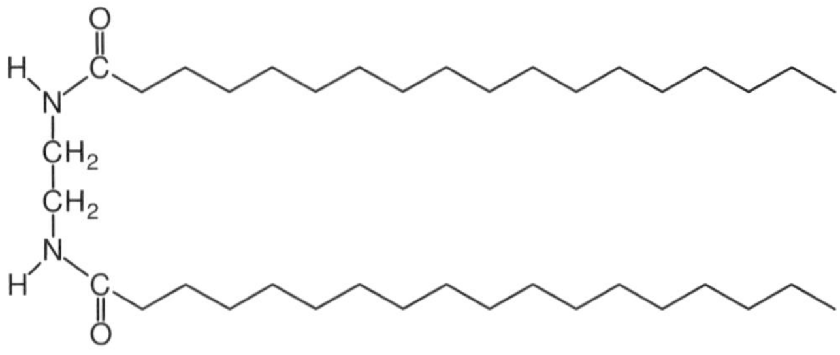

90   Additives for Thermoplastics

Figure 8.2: Bis(stearoyl)ethylenediamine

8.3.2   Fatty Acids, Esters and Amides

There is a considerable overlap in the action of phthalic acid esters. They act in several ways, namely, as plasticizers, lubricants, and mold release agents (5).

Fatty acids show a good lubricating effect in PVC. Fatty esters from short chain alcohols are important lubricating additives for PVC.

The diamide from ethylenediamine, bis(stearoyl)ethylenediamine is commonly known as amide wax. The structure is shown in Figure 8.2 and it is a highly versatile lubricant.

Stearamide-based lubricants have been used in compositions based on acrylonitrile-butadiene-styrene (2).

8.3.3   Waxes

Montan wax is a cheap lubricant. It is obtained as by-product from the production of lignites. Commercially interesting deposits exist in only a few locations of the world, e.g., near Amsdorf, Germany and near Ione, California.

Montan wax is extracted by solvent extraction from lignite. From the chemical perspective, montan wax is somewhat related to fatty acids. It can be bleached by treating with hot chromo-sulfuric acid.

Polar poly(ethylene) waxes and polar poly(propylene) waxes are obtained by the oxidative treatment of the respective polymers. After treatment, these materials then bear oxygen-containing groups, either hydroxy groups or oxo groups, in their side chains, or at the end.

Compositions for the production of sintered molded parts are composed of a metallic, ceramic, and polymer material and a compaction aid, The latter is a mixture of a poly(glycol), and a montan wax (3).

8.3.4 Polymeric Lubricants

Poly(tetrafluoroethylene) polymers have found a limited use as lubricants. They arse used for materials that should exhibit a high abrasion resistance.

8.4 Special Applications

8.4.1 PVC

In general, a PVC resin contains aids such as a lubricant and a processing aid, in order to improve its moldability and processability and to enable it to stably pass through various processes. Polymers such as a copolymer mainly composed of methyl methacrylate have been used as such polymer processing aids or lubricants (6,7).

8.4.2 Chlorinated PVC

In chlorinated PVC, the high content of chlorine imparts some favorable characteristics, but also confers other characteristics that are not as desirable, i.e., the rigidity of the molecule impedes its movement and thus it is a material which is more difficult to process, and at the same time it becomes fragile and, as PVC, tends to breakdown under the high temperature conditions required for its processing.

For this reason, chlorinated PVC has to be mixed with several chemical agents in order to retain the desired behavior, both in the processing stage as well as during its performance as a product. Among the components added to it, thermal stabilizers prevent and delay degradation, impact modifiers make it less fragile, and process auxiliaries as well as lubricants among other additives obtain a good property balance.

Examples of lubricants for chlorinated PVC are metal stearates, montan waxes, high and low molecular weight paraffinic waxes, as well as oxidized waxes. Specific examples of external lubricants are Advalube® E-2100 available from Rohm and Haas, Licolub® XL 445 available from Clariant. These lubricants are used at levels ranging from 4–5.5 phr (8).

8.4.3 Electrically Conductive Polymers

In electrically conductive poly(oxymethylene) (POM) molding compositions special problems with the lubricant composition emerge (9). Only a few lubricants are capable of reducing abrasion when used with conductivity black incorporated into POM. It is likely that they disrupt the bonding between matrix and carbon black. This may well also be the reason for the reduction in mechanical properties. Therefore, careful selection and use of the lubricants is required.

The electrical conductivity is maintained by the addition of a lubricant mixture composed of a lubricant with predominantly external lubricant action and of a lubricant with predominantly internal lubricant action.

Preference to the former type has been given to a high-molecular-weight, oxidized and therefore polar polyethylene wax. This improves the tribological performance and can reduce the severity of drop down in mechanical properties. Stearyl stearate is preferably used as an internal lubricant that establishes gentle conditions for the incorporation of the carbon black.

References

1. R. Levy, Lubricant compositions and methods, US Patent 7 358 216, assigned to Lee County Mosquito Control District (Fort Myers, FL), April 15, 2008.
2. J. Lee, S.H. Chae, J. Cha, S. Kim, and C. Lee, Plastic resin composition having improved heat resistance, weld strength, chemical resistance, impact strength, elongation, and wettability, US Patent 7 557 159, assigned to LG Chem, Ltd. (KR), July 7, 2009.

3. R. Lindenau, K. Dollmeier, and V. Arnhold, Composition for the production of sintered molded parts, US Patent 7 524 352, assigned to GKNM Sinter Metals GmbH (Radevormwald, DE), April 28, 2009.
4. B.B. Wilson, R.J. Stumo, S.C. Erickson, W.L. Kopecky, and J.C. Breister, Lubricated flow fiber extrusion, US Patent 7 476 352, assigned to 3M Innovative Properties Company (St. Paul, MN), January 13, 2009.
5. N. Agarwal, S.K. Gaggar, D. Gupta, S. Gupta, R. Krishnamurthy, N. Preschilla, R.S. Totad, and S. Tyagi, Polymer compositions, method of manufacture, and articles formed therefrom, US Patent 7 557 154, assigned to SABIC Innovative Plastics IP B.V. (Bergen op Zoom, NL), July 7, 2009.
6. K. Nakamura and A. Nakata, Lubricant for a thermoplastic resin and a thermoplastic resin composition comprising thereof, US Patent 5 942 581, assigned to Mitsubishi Rayon Co., Ltd. (Tokyo, JP), August 24, 1999.
7. K. Nakamura and M. Ito, Impact modifier, process for producing the same, and thermoplastic resin composition containing the same, US Patent 6 833 409, assigned to Mitsubishi Rayon Co., Ltd. (Tokyo, JP), December 21, 2004.
8. O.P. Tabla, L.V. Estrada, and A.P. Sanchez, Thermoplastic formulations for manufacturing pipes and accessories for home and industrial use, and process for the same, US Patent 7 423 081, assigned to Servicios Condumex S.A. de C.V. (Queretaro QRO, MX), September 9, 2008.
9. O. Schleith and K. Kurz, Slip-modified, electrically conductive polyoxymethylene, US Patent 6 790 385, assigned to Ticona GmbH (DE), September 14, 2004.

9
Antistatic Additives

Many polymers exhibit good electric insulation properties and this makes them prone to accumulate electric charge. Static electricity is generated by the close contact of two materials of different composition, e.g., by rubbing. Dryness is an important factor for the generation of static electricity. Therefore, in winter time the problem of static electricity is still more pronounced.

For the reasons pointed out above, there is a need of adding antistatic agents to polymers. The field of application of antistatic additives include (1, pp. 213):

- Packaging: Reduction of dust accumulation on consumer goods
- Electronics: Avoiding short circuits
- Industry: General precaution
- Processing of plastics: Avoiding bridging during feeding.

Many antistatic additives are either liquid at room temperature or melt at temperatures below the processing temperature. Often liquid antistatic additives are sold in solid form, as they are absorbed on a carrier such as silica, or they are in the form of a highly concentrated batch.

Special precautions are required for applications in food packaging, cosmetic products, products for household, etc.

9.1 Types of Additives

The antistatic properties of the polymers depend on the ambient humidity and they are not permanent, since these agents migrate to the surface of the polymer and disappear. Polymers or oligomers were then provided as antistatic agents. These agents have the advantage of not migrating and of giving permanent antistatic properties, which are, moreover, independent of the ambient humidity (2).

High molecular weight additives are migrating slower than low molecular weight types. For this reason the former retain their efficiency for a longer time (3, p. 53).

The incorporation of antistatic additives increases either the surface conductivity or the volume conductivity. Therefore, there are two basic types, namely, internal and external antistatic additives (4, p. 26). External antistatic additives are dissolved in appropriate solvent and this solution is sprayed or coated onto the polymer.

Internal antistatic additives are incorporated into the polymeric matrix. The additives migrate onto the surface of the polymer and they function by two mechanisms:

1. They act as lubricants and reduce friction which cause charging or
2. They are providing increased electric conductivity of the surface.

The issues of internal and external additives are summarized in Table 9.1.

9.2 Areas of Application

The largest quantities of additives are used in the field of poly(ethylene) (PE) and poly(propylene) (PP). Applications are typically in packaging. Other polymers where antistatic additives are used are poly(styrene) (PS) and poly(vinyl chloride) (PVC) (3). PVC requires special consideration, because there are two types, i.e., rigid PVC and soft PVC. The proper selection of antistatic additives has been extensively documented (5). Among engineering plastics, antistatic additives did not find widespread use. This arises because (3):

1. The thermal stability is not sufficient to survive the high processing temperatures, and
2. The additive caused undesired side effects, such as coloring

Table 9.1: Comparison of Internal and External Additives (6, p. 631)

Issue	Internal Additive	External Additive
Dose	0.1–3%	0.1–10%
Application	Singularly or together with other additives	Coated by secondary processes
Durability	Maybe removed from surface by washing or abrasion but delivered from interior	Performance lost if washed out
Time	Due to migration it needs some time to become effective	Acts instantly
Types	Must be in accordance with polymer type	Usable regardless of the polymer type
Equipment	No additional equipment necessary	Coating equipment is needed
Precautions	Possible sensitive to high temperature processing	Some surfaces are hard to be wetted
	Degradation may influence clarity	Uniform coating required
	Printing, heat sealing may be problematic	Protective coating equipment, e.g., ventilating is necessary
	Testing the process is necessary	Some additives may initiate corrosion

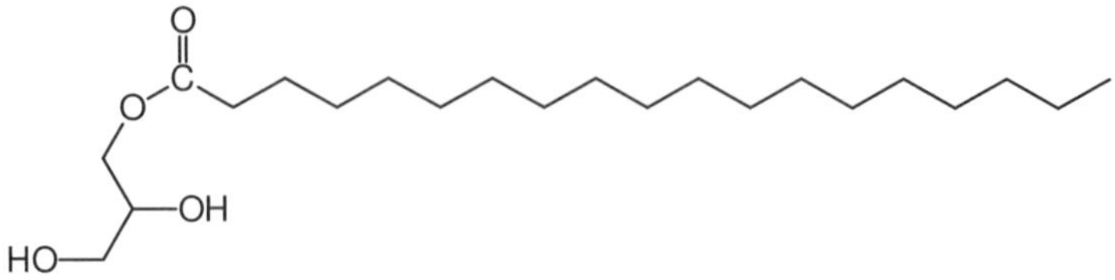

Figure 9.1: Glycerol monostearate

of transparent materials, for example, poly(methyl methacrylate) (PMMA) or poly(carbonate) (PC).

Cationic additives are most suitable in polar polymers, such as PVC and styrenics. The performance of anionic additives is comparable to cationic additives. Non-ionic additives are less polar than ionic additives. They are used in PE and friends.

9.3 Additives in Detail

9.3.1 Conventional Additives

Well known antistatic agents, which are added to polymers are ionic surfactants of the ethoxylated amine and sulfonate types.

Ethoxylated amines have the basic structure like

$$R-N(CH_2-CH_2-OH)_2.$$

They are used in PE, PP, and acrylonitrile-butadiene-styrene (ABS).

Glycerol monostearate is used mainly in PP. The structure is shown in Figure 9.1.

Fatty alcohol amides are used in PE and PP. These types act for a prolonged time. Amide based types are composed from lauric acid $CH_3-(CH_2)_{10}-COOH$ and cocoa fatty acid.

Sodium alkyl sulfonates are used for PVC and PS types. They are suitable for high temperature processing.

Typical classes of additives are shown in Figure 9.2.

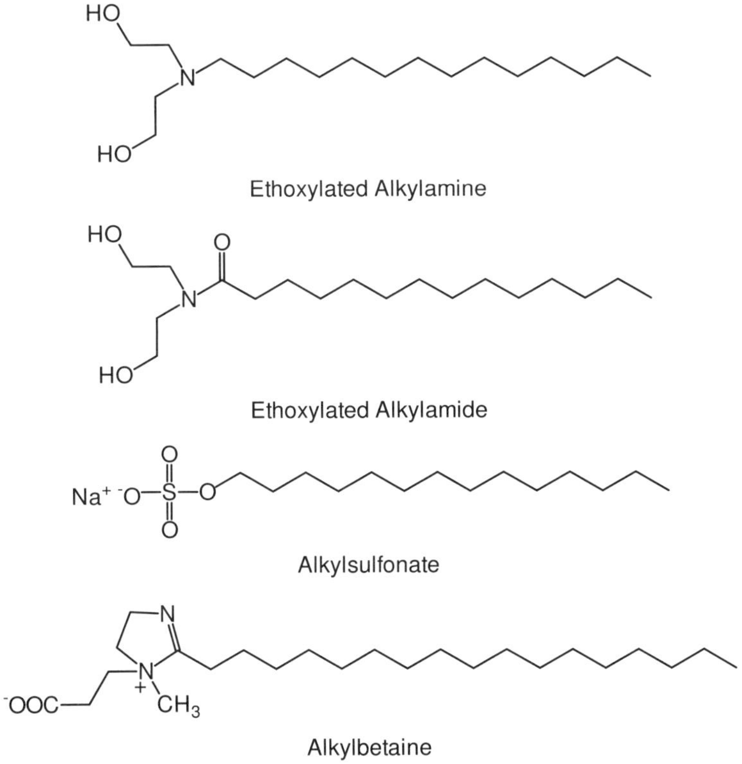

Figure 9.2: Classes of Antistatic Additives (6, pp. 636)

Table 9.2: Properties of a Polymeric Additive Composition (7)

Composition					
Ratio	75:25:0	72:28:0	39:13:23	70:30:0	39:13:23
Styrenic[a]	70	70	70	75	85
Epichlorohydrin[a]	15	15	15	18	15
PMMA[a]	15	15	15	7	–
Properties					
Surface resistivity[b]	1E12	1E12	1E12	1E12	1E12
% Tensile elongation[c]	20–25	30–40	40–50	19	2–4
Dissipation[d,e]	1.37	1.62	1.12	0.77	2.34
Dissipation[d,f]	9.00	11.00	3.00	1.24	5.00

[a] Polymer, parts by weight
[b] ASTM D-257
[c] ASTM D-638
[d] Federal Test Method Standard 101B, Method 4046
[e] Electrostatic dissipation sec. at 50% relative humidity
[f] Electrostatic dissipation sec. at 15% relative humidity

9.3.2 Polymeric Additives

Antistatic compositions are made from poly(ether ester amide) and a thermoplastic resin, functionalized by acrylic acid or maleic anhydride (8).

PVC, PE, PP, ABS resins, and PS, can be rendered antistatic by adding an antistatic blend composed of a copolymer comprising poly(amide) blocks and poly(ether) blocks, sodium perchlorate, and of a fibrous material (9).

Styrene acrylonitrile copolymers can be rendered to be antistatic by adding a copolymer composed from epichlorohydrin and ethylene oxide and PMMA (7). The mixture is compounded in a Banbury mixer and then injection molded at 220°C. The surface resistivity, the percent of tensile elongation and the electrostatic dissipative rate from 5 kV to 0 V at 15 and 50% relative humidity have been measured. The results are given Table 9.2.

9.3.3 External Antistatic Additives

External antistatic additives are sometimes addressed as topical antistatic additives because they are essentially applied as a coating on the plastic surface. The method of coating requires only a relatively small amount of additive which acts immediately. However, the antistatic properties are are less permanent. For example, the specific properties may be easily lost when the surface is wiped or otherwise cleaned.

External antistatic additives are mostly used for fibers, e.g., PVC floor finishing. Usually the article is sprayed with a solution containing the additive. It is important to emphasize that 3,4-poly(ethylenedioxythiophene) belongs to the class of conductive coatings (10).

9.3.4 Intrinsically Antistatic Compositions

Antistatic thermoplastic resin compositions that are intrinsically antistatic have been described. These are, for example, thermoplastic resin compositions composed from poly(ether ester amide) and a thermoplastic resin e.g., styrene based resins, poly(phenylene ether) resins and poly(carbonate) resins. These compositions are permanently antistatic and are excellent in their mechanical properties. The compositions are suitable for housing of optical or magnetic recording media (11).

Compositions made from poly(ether ester amide)s, styrene resins, poly(phenylene oxide) and PC are suitable antistatic compounds. Preferably, the compositions also comprise PS modified by methyl acrylate (11). Copolymers comprising polymer blocks and poly(ether) blocks $-C_2H_4-O-$ need a compatibilizing agent. This is chosen from copolymers of styrene and maleic anhydride (12).

9.3.5 Conductive Fillers

Conductive fillers and fibers are well established and here has been little development in the recent years.

The electrical conductivity does not vary linearly with the amount of loading, but, there is a critical loading where the conductivity increases abruptly. The critical loading is also addressed as percolation threshold.

Table 9.3: Properties of Stainless Steel Fibers (3, p. 62)

Property	ABS	PC	PPO	PA 6,6
Loading/[%]	7	7	7	7
Density/[g cm^{-3}]	1.12	1.27	1.18	1.22
Tensile strength/[psi]	9,500	9,300	7,800	8,500
Flexural strength/[psi]	13,500	14,000	13,700	17,000
Flexural modulus/[psi]	0.45E6	0.38E6	0.33E6	0.64E6
Attenuation at 1 GHz	40	40–50	35–40	35–40

The critical loading is dependent on the matrix material, the particle size distribution, and the aspect ratio. The aspect ratio is the ratio of length to width of the particle. For example, fibers exhibit a high aspect ratio.

For carbon particles, the theory predicts that for a given volume of loading, a large number of small particles is more effective than a small number of big particles. In general, irregularly shaped particles are more effective than spherical particles.

Various theories have been developed in order to explain the electrical conductivity of polymers filled with carbon black, These include (3, p. 59):

1. Conductivity path theory: The formation of continuous chains made up from carbon particles, and
2. Electrical conductivity occurs by jumping of electrons across gaps of carbon particles in the polymeric matrix.

Experience shows that the theory can predict the observed phenomena very well. Besides carbon particles there are other conductive fillers, such as poly(acrylonitrile) carbon fibers, pitch carbon fibers, graphite fibers, metal fibers and metal coated organic fibers.

Stainless steel fibers are very fine, i.e. in the range of 10 μ. They exhibit a very high flexibility.

Typical properties of stainless steel fibers are given in Table 9.3.

Silver coated glass spheres are used in special electronic applications for conductively fixing electronic parts. Thus they are used in thermosetting resins. Typical loading levels are in the range of 60–65%.

Much attention has been given to the use of carbon nanotubes

(CNT)s as filler in conductive nanocomposites in order to harness their exceptional electrical properties (13, 14). CNTs can be used as the dispersed conductive phase in an insulating polymer matrix. They can be used at comparatively low loadings which reduce weight. Further, the interference with other polymer properties is minimized (15). The largest field of applications is the automotive industries, such as electrically conductive polyamide fuel lines.

With respect to their electrical conductivity, nanocomposites exhibit a percolation threshold between 0.5–1%. Below 1% loading the nanocomposites exhibit properties that are very similar to the resin matrix, i.e., acting as an insulator. Higher loadings change the electric properties into conductors. A maximum conductance of 0.06 S/cm at 4% loading can be obtained. The majority of the research in nanocomposites targets to the improvement of their mechanical properties. This is justified, because the nanotubes increase the stiffness and strength of the pure materials (16).

Graphene is something like an exfoliated graphite. It can be used in optically transparent materials. Graphene is claimed to be still superior in comparison to CNTs. Graphene based polymer composites have potential applications in radiation and electromagnetic shielding, antistatic, shrinkage resistant and corrosion resistant coatings (16).

In the case of conductive nanocomposites, the target is to obtain a network of connected filler particles which allows electrical current flow through the sample. The advantages of both types of materials, namely, the high conductivity of the CNTs and the good processability of polymeric materials can be combined (17).

It has been found that the percolation threshold is influenced by the processing temperature. The percolation threshold is in the range of 0.9–1.3% loading of CNT. The electrical conductivity against the amount of filler is shown in Figure 9.3.

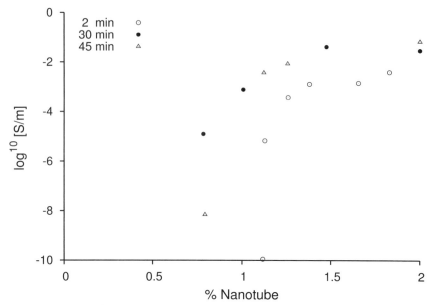

Figure 9.3: Conductivity Measurements of Multiwalled Carbon Nanotube Poly(styrene) Composites Pressed at 125°C and 100 bar for Several Processing Times (17)

References

1. A. Müller, *Coloring of plastics. Fundamentals, colorants, preparations*, Carl Hanser Verlag, Munich, Vienna, New York, 2003.
2. C. Lacroix, R. Linemann, P. Blondel, and Y. Lermat, Antistatic polymer compositions, US Patent 6 913 804, assigned to Arkema (Paris la Defense Cedex, FR), July 5, 2005.
3. N. Drake, *Polymeric Materials for Electrostatic Applications: Industry Analysis Report*, Rapra Technology Ltd., Shawbury, 1996.
4. J. Leadbitter, J.A. Day, and J.L. Ryan, *PVC: Compounds, Processing and Applications*, Rapra Technology Ltd., Shawbury, 1994.
5. J. Pionteck and G. Wypych, eds., *Handbook of Antistatics*, Vol. 2, ChemTec Publ., Toronto, 1st edition, 2007.
6. H. Zweifel, ed., *Plastics Additives Handbook*, Hanser Publishers, Munich, 5th edition, 2001.
7. S.K. Gaggar, J.M. Dumler, and T.B. Cleveland, Polymer blend compositions, US Patent 4 857 590, assigned to GE Chemicals, Inc. (Parkersburg, WV), August 15, 1989.

8. M. Tanaka, A. Kishimoto, and T. Hirai, Antistatic resin composition, JP Patent 60 023 435, assigned to Toray Industries, February 06, 1985.
9. B. Hilti, E. Minder, J. Pfeiffer, and M. Grob, Antistatic composition, EP Patent 0 829 520, assigned to Ciba Geigy AG, March 18, 1998.
10. F. Jonas and J.T. Morrison, 3,4-Polyethylenedioxythiophene (PEDT): Conductive coatings technical applications and properties, *Synthetic Metals*, 85(1-3):1397–1398, 1997.
11. T. Fukumoto, M. Iwamoto, and A. Kishimoto, Intrinsically antistatic thermoplastic resin compositions., EP Patent 0 242 158, assigned to Toray Industries, October 21, 1987.
12. C. Lacroix, Antistatic styrene polymer compositions, WO Patent 0 110 951, assigned to Atofina and Lacroix Christophe, February 15, 2001.
13. J.W.G. Wildoer, L.C. Venema, A.G. Rinzler, R.E. Smalley, and C. Dekker, Electronic structure of carbon nanotubes investigated by scanning tunneling spectroscopy, *Nature*, 391:59–62, 1998.
14. T.W. Odom, J.L. Huang, P. Kim, and C.M. Lieber, Atomic structure and electronic properties of single-walled carbon nanotubes, *Nature*, 391 (6662):62–64, 1998.
15. J. Markarian, New developments in antistatic and conductive additives, *Plastics, Additives and Compounding*, 10(5):22–25, 0 2008.
16. B. Debelak and K. Lafdi, Use of exfoliated graphite filler to enhance polymer physical properties, *Carbon*, 45(9):1727–1734, August 2007.
17. N. Grossiord, P.J.J. Kivit, J. Loos, J. Meuldijk, A.V. Kyrylyuk, P. van der Schoot, and C.E. Koning, On the influence of the processing conditions on the performance of electrically conductive carbon nanotube/polymer nanocomposites, *Polymer*, 49(12):2866–2872, June 2008.

10
Slip Agents

Due to their high coefficient of friction, poly(olefin) films tend to adhere either mutually or to the production equipment during processing.

Slip additives act as they modify the surface properties of the polymeric materials. They reduce the friction coefficient of any other surfaces with which they come into contact. Thus slip agents (1):

1. Facilitate an increased line speed in the manufacturing process, when applied as processing aids, and
2. Eventually enhance the performance of the packaging machine by reducing the coefficient of friction.

Slip agents are sometimes referred to as lubricants. However, it has been pointed out that there is a difference in between slip agents and lubricants, in as much as lubricants are intended to be applied exclusively as processing aids (2). Most slip agents can serve as processing aids, but the reverse is not true in general. This curious situation arises, because not all slip agents function as external agent and so they may not act immediately after incorporation into the polymeric matrix. Admittedly, this strict definition is not in general use in the literature.

10.1 Basic Principles of Action

Slipping describes the sliding of parallel film surfaces over each other or the sliding of film surfaces over substrates (3).

The slip agents are added to poly(ethylene) (PE) or poly(propyl-

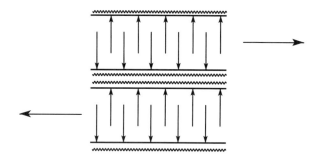

Figure 10.1: Basic Principle of Action of a Slip Agent

ene) (PP) films during the extrusion process to decrease friction, i.e., film-to-film friction and film-to-production equipment.

In this way, the output of the machine can be increased. Further secondary operations, including packaging operations, can be accelerated. The slip effect is measured in terms of the coefficient of friction.

A decreased coefficient of friction results from the migration of slip additive to the film surface. This arises due from incompatibility of the slip additive with the polymer.

The migration rate of slip agent to the surface is largely caused by the chain length of the additive. Namely, the chain length is related to the compatibility with the polymer.

The basic principle of action is shown in Figure 10.1.

Further, the crystallinity of the polymer plays a role: larger the slip agent the more compatible it is with the polymer and the slower the migration takes place. For example, oleamides migrate faster than erucamides.

However, erucamides are more heat stable than oleamides, more resistant to oxidation, and the former create fewer volatiles during processing.

Thus, erucamides are more suitable for higher processing temperatures and processes with high output, which will result in high quality final products. PE is less crystalline than PP. For this reason the migration of the slip additive is more pronounced in PE.

Actually, because of their general required properties, slip agents are not only used as such, but are also suitable for use as demold-

Table 10.1: Slip Agents (4)

Compound
Saturated fatty acid amides
Stearamide Behenamide Ethylene bis stearamide
Unsaturated fatty acid amides
Oleamide Erucamide Linoleamide Ethylene bis oleamide Stearyl erucamide
Poly(ether) poly(ol)s
Carbowax
Fluoro-containing Polymers
Poly(tetrafluoroethylene) Fluorine oils and waxes
Silicon Compounds
Silanes Silicone polymers

ing agents in injection moulding techniques. Moreover, slip agents function as antiblocking agents as well.

10.2 Compounds

As previously mentioned, the most commonly used slip agents are oleamides and erucamides. Slip agents are summarized in Table 10.1. Selected slip agents are shown in Figure 10.2.

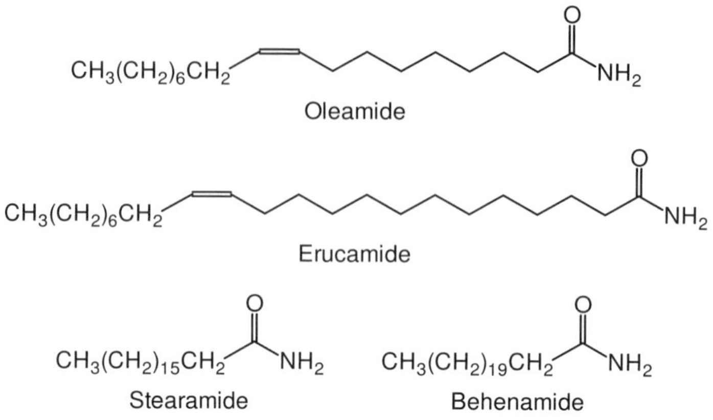

Oleamide, Erucamide, Stearamide, Behenamide structures

CH₃(CH₂)₆CH₂—...—C(=O)NH₂
Oleamide

CH₃(CH₂)₆CH₂—...—C(=O)NH₂
Erucamide

CH₃(CH₂)₁₅CH₂—C(=O)NH₂
Stearamide

CH₃(CH₂)₁₉CH₂—C(=O)NH₂
Behenamide

Figure 10.2: Slip Agents

10.3 Special Formulations

10.3.1 Poly(ethylene terephthalate)

Poly(ethylene terephthalate) (PET) is often used for beverage containers. For such applications, the fabricated PET bottle must exhibit low color and high transparency, and in addition to low taste and odor, the additive must be non-toxic. These requirements impose further important issues for a slip agent in addition to its friction-reducing properties (5).

Conventional amide slip agents, e.g., erucamide, lower the coefficient of friction to 62–71% of the blank material, but result in severe yellowing of the polymer. In addition, the effect is short lived, and after one week the additive effect disappears.

Additives conforming to this criteria afford an equivalent or greater reduction in the coefficient of friction when compared with conventional amide slip agents, but the polymer plaques remain clear and transparent. Using stearyl behenate and stearyl palmitate as slip agent, very low coefficients of friction of 30–50% of the blank can be achieved, at levels of only 0.2–0.3% (5).

10.3.2 Formulations for Poly(ethylene)

Masterbatches for PE have been described that contain a clarifying agent, a high clarity antiblock and an amide slip additive (6). These masterbatches can be used for the production of articles from both low density poly(ethylene) and linear low density poly(ethylene) by either a cast, blown or molding process. The articles show an increased gloss, a reduced haze, a reduced coefficient of friction, as well a reduced blocking force. The slip agent is erucamide.

References

1. SpecialChem S.A., Why to use slip agents?, [electronic:] http://www.specialchem4polymers.com/tc/Slip-Agents/index.aspx, 2009.
2. C.A. Harper and E.M. Petrie, eds., *Plastics Materials and Processes: A Concise Encyclopedia*, John Wiley and Sons, New York, 2003.
3. G. Wypych, *Handbook of Antiblocking, Release, and Slip Additives*, Vol. 1 of *Encyclopedia of Polymer Additives*, ChemTec Publ., Toronto, 1st edition, 2005.
4. A.K. Mehta, S. Datta, W. Li, and S.S. Iyer, Heterogeneous polymer blends and molded articles therefrom, US Patent 7 476 710, assigned to ExxonMobil Chemical Patents Inc. (Houston, TX), January 13, 2009.
5. D.A. Parker, A. Maltby, M. Read, and P. McCoy, Aliphatic ester compounds as slip agents in polyester polymers, US Patent 7 501 467, assigned to Croda International PLC (North Humberside, GB), March 10, 2009.
6. D. Beuke, P. Pickett, P. Patel, S. Lucas, P. Trivedi, and N. Savargaonkar, Polyethylene formulations, US Patent 7 365 117, assigned to Ampacet Corporation (Tarrytown, NY), April 29, 2008.

11
Surface Improvers

This class of compounds is also unspecifically addressed as polymer processing aids.

In the manufacture of extruded polymers there are a number of surface defects referred to as sharkskin, snakeskin and orange peel which all are related to the rheology of the polymer melt and in particular the melt fracture of the polymer (1).

Melt fracture is a flow phenomenon that occurs as the molten polymer flows through the die, starting at the die entry, evidenced by gross irregularities in the shape or surface of the extrudate. Melt fracture is considered to be the result of non-uniform or irregular elastic strains in the material at the die entrance. The shear rate at the surface of the polymer is sufficiently high that the surface of the polymer begins to fracture. That is, there is a slippage of the surface of the extruded polymer relative to the body of the polymer melt. The surface generally cannot flow fast enough to keep up with the body of the extrudate and a fracture in the melt occurs. These irregularities in the shape or surface of the extrudate are undesirable. For example, the irregularities produce an unattractive pattern on blown films. Processing aids are typically added to the polymer so that during melt processing they will migrate to the surface of the polymer lubricating the polymer and die surfaces thereby allowing high throughput with reduced melt fracture (2).

However, it should be emphasized that melt fracture is not always considered as a draw back:

> Beautifying patterns on the surface of extruded polymer products can be produced by forcing molten polymer through a die which is coated with a con-

trolled pattern of low surface energy material. Melt fracture of the poly(ethylene) as it emerges from the die is produced at gaps in the coating material which may be silicon, inorganic or fluorine containing polymers (3).

11.1 Additives

11.1.1 Fluorocarbon Compounds

Certain fluorocarbon processing aids are known to partially alleviate melt defects in extrudable thermoplastic hydrocarbon polymers and allow for faster, more efficient extrusion. Blatz first described the use of fluorocarbon polymer process aids with melt extrudable hydrocarbon polymers wherein the fluorinated polymers are homopolymers and copolymers of fluorinated olefins having an atomic fluorine to carbon ratio of at least 1:2, wherein the fluorocarbon polymers have melt flow characteristics similar to that of the hydrocarbon polymers (4).

Vinylidene fluoride homopolymers can be used to improve the extrusion of low density poly(ethylene) (5).

Fluoroelastomers are useful as processing aids. The fluoropolymers are normally in the fluid state at room temperature (2). The effect of the addition of an elastomer to linear low density poly(ethylene) (LLDPE) is shown in Figure 11.1.

Multimodal fluoropolymers are superior in performance than unimodal fluoroplastics when used as polymer processing additive (6).

11.1.2 Acrylics

In addition an acrylic based processing aid improves the extrusion of high density poly(ethylene) (HDPE). However, the use of this processing aid provides only a marginal improvement in the extrusion of LLDPE having a narrow molecular weight distribution (7).

Acrylic polymers are generally much less expensive than fluoropolymers. Accordingly, there is an economic incentive to employ acrylics rather than fluoropolymers as processing additives. How-

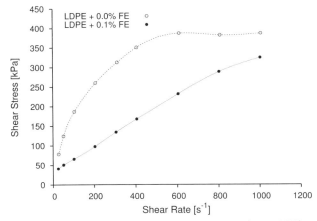

Figure 11.1: Effect of Fluoroelastomer (8, p. 570)

ever, the simple addition of an acrylic polymer and a fluoropolymer to a poly(ethylene) extrusion process has been observed to produce antagonistic results, i.e., the performance produced by one has been observed to be adversely affected by the simple addition of the other (1).

Homogeneous blends of a thermoplastic styrene/methyl methacrylate (SMMA) copolymer and a thermoplastic vinylidene fluoride/hexafluoropropylene copolymer which is an excellent processing additive for the extrusion of thermoplastic polyolefin (1).

11.1.3 Modified Pigments

A surface treated titanium dioxide pigment together with a polymer processing aid improves the effectiveness of the polymer processing aid in reducing melt fracture. Only a single polymer masterbatch, containing both pigment and polymer processing aid is needed.

The titanium dioxide pigment is surface treated with methylhydridosiloxanes. These have the formula

$$(CH_3)_3SiO[SiOCH_3)H]_n-SiCH_3)_3.$$

Typically, n is ranging from about 30 to about 70. The surface treated pigment resulting from the pigment coming into contact with the silicon-containing compound, comes in contact with a second surface treatment of an organic compound selected from a group consisting of a hydrocarbon wax, a carboxylic acid, and a silicone polymer. This mixture forms a pigment having a silicon-containing surface treatment and an organic surface treatment. The surface treated pigment is dispersible throughout the polymer melt and promotes migration of the processing aid to a surface of the polymer melt (2).

11.1.4 Organic Salts

Zinc stearate has been demonstrated to reduce the melt fracture in poly(ester)s (9). Zinc salts may be mixed with the poly(ester) in a wide range of melt processing equipment such as Banbury mixers or extruders. Either single screw or twin screw extruders may be used.

A preferred process of adding the zinc salt to the poly(ester) is to make a masterbatch of the zinc salt of an organic acid in a polymer base material such as a poly(ester) or a poly(olefin). Suitable poly(olefin)s include poly(propylene) and LLDPE, or HDPE. The masterbatch is mixed in the course of the extrusion operation into the desired poly(ester). The concentration of zinc salt in the masterbatch may range from 5–30%.

References

1. P.S. Chisholm, T. Tikuisis, S.K. Goyal, D. Checknita, and N.K.K. Bohnet, Melt fracture reduction, US Patent 5 854 352, assigned to Nova Chemical Ltd. (Calgary, CA), December 29, 1998.
2. D.D. May and B. Zimmermann, Titanium dioxide–containing polymers and films with reduced melt fracture, US Patent 7 338 995, assigned to E.I. du Pont de Nemours and Company (Wilmington, DE), March 4, 2008.
3. T.K. Su, Method and apparatus producing decorative melt fracture patterns on polymeric products, US Patent 4 615 858, assigned to Mobil Oil Corporation (New York, NY), October 7, 1986.
4. P.S. Blatz, Extrudable composition consisting of a polyolefin and a fluorocarbon polymer, US Patent 3 125 547, assigned to E.I. DuPont de Nemours and Co., March 17, 1964.
5. S.C. Chu and R.G. Shaw, Compositions of linear polyethylene and polyvinylidene fluoride for film extrusion, and films thereof, US Patent

4 753 995, assigned to Mobil Oil Corporation (New York, NY), June 28, 1988.
6. M.P. Dillon, S.S. Woods, K.J. Fronek, C. Lavallee, S.E. Amos, K.D. Weilandt, H. Kaspar, B. Hirsch, K. Hintzer, and P.J. Scott, Polymer processing additive containing a multimodal fluoropolymer and melt processable thermoplastic polymer composition employing the same, US Patent 6 277 919, assigned to Dyneon LLC (Oakdale, MN), August 21, 2001.
7. W.D. Heitz, Paraloid extrusion aids for high molecular weight hdpe film resins, US Patent 4 963 622, assigned to Union Carbide Chemicals and Plastics Company Inc. (Danbury, CT), October 16, 1990.
8. H. Zweifel, ed., *Plastics Additives Handbook*, Hanser Publishers, Munich, 5th edition, 2001.
9. D.C. Cobb and J.W. Mercer, Jr., Process for making thermoplastic profiles having reduced melt fracture, US Patent 6 100 320, assigned to Eastman Chemical Company (Kingsport, TN), August 8, 2000.

12

Nucleating Agents

A nucleating agent is defined to be an additive which forms nuclei in a polymer melt to promote the growth of crystals (1). In addition, a nucleating agent is often a clarifying agent, or clarity-enhancing agent, respectively. The issues of nucleating agents have been reviewed in the literature (2).

The crystallization of polymers is controlled by the nucleation stage, and the addition of specific additives or nucleating agents to shorten the induction time of crystallization and accelerate the formation of crystalline nuclei is a technique commonly used in the polymer industry to shorten injection moulding cycle times, thus reducing production costs. Furthermore, such agents generate smaller spherulites and increase crystallinity, thus improving the optical and mechanical properties (3).

Actually, there is a difference in between nucleation rate and crystallization rate. Nucleation refers to the formation of the smallest entities that may subsequently form crystals. Crystallization accelerators refer to additives that promote the secondary steps of crystal growth, once nuclei are formed.

Clarity-enhancing agents provide a site within the polymer resin where crystallization may occur. This allows smaller crystallites to form. The smaller crystallites allow light to travel through the polymer medium and avoid diffraction of the light (4). The transparency of plastics can be measured according to a standard test method (5).

12.1 Crystalline Polymers

Some crystalline polymers, e.g., polyolefin polymers such as poly(ethylene), poly(propylene) (PP) and poly(1-butene), and poly(ester) polymers such as poly(ethylene terephthalate), and poly(amide) (PA) polymers have a slow rate of crystallization after heat forming. Consequently, the molding cycle when they are processed is too long, and as crystallization proceeds even after molding is complete, the molded product is sometimes deformed. In addition, there is the disadvantage that these crystalline polymer compound materials formed large spheroid crystals, so their mechanical strength and transparency is poor.

These disadvantages are due to the crystalline properties of the crystalline polymer compounds. The solution is to form fine crystals very rapidly. In order to do this, the crystallization temperature may be increased, or crystal nucleating agents and crystallization promoters may be added (6).

12.1.1 Crystal Structures

Polymers may crystallize in different structures, for example, isotactic PP is known to crystallize in α, β, γ and smectic structures (7). Nucleating agents may favor specific types of crystallites. For isotactic PP, stearic acid and stearate lanthanum complexes are β-nucleating agents. Similarly, cadmium bicyclo[2.2.1]hept-5-ene-2,3-dicarboxylate is a β-nucleating agent (8).

12.1.2 Modification of Properties by Crystallinity

A method of modifying the properties of PP, either as a homopolymer or as a copolymer, is to modify its crystalline structure. The onset of crystallinity is called nucleation and this may occur randomly throughout the polymer matrix as the individual polymer molecules begin to align.

Alternatively, nucleation may occur at the interface of a foreign impurity or an intentionally added nucleating agent. The proper use of nucleating agents can result not only in unique and desirable crystalline structures, but they can also promote the efficiency of a given process by shortening process times, or by initiating nucleation at higher temperatures (9).

12.2 Experimental Methods

12.2.1 Nucleation Technologies

Nucleation technologies can be subdivided into (10):

- Conventional
- Advanced
- Hyper nucleation.

Hyper nucleating agents have been developed recently and offer the combined benefits of high crystallization rates and isotropic shrinkage control, which leads to improved production and part quality performance, in addition to the traditional mechanical property-related enhancements. Hyper nucleating agents are formulated compounds. For example, Hyperform® HPN-68L is a blend of the disodium salt of bicyclo[2.2.1]heptane-2,3-dicarboxylic acid, Erucamide and amorphous silicon dioxide (11).

12.2.2 Characterization of Polymer Crystallization

The mechanism about polymer crystallization in the presence of nucleating agent is complicated. Nucleated crystallization behaviors have been widely studied through traditional techniques, including differential scanning calorimetry (DSC), polarized light microscope, X-ray scattering, and rheological techniques (12).

Typically, dynamic crystallization experiments are carried out by cooling down to subambient temperature at a cooling rate of 10 Kmin^{-1}, after melting the samples and keeping them above melt temperature in order to erase the thermal history.

The transition temperatures are simply taken as the peak maxima in the calorimetric curves. In blends of isotactic PP, the crystallinity X_c is determined from the crystallization exotherms by using the following relation (3):

$$X_c = \frac{\Delta H_c}{\Delta H_c^0 w}, \qquad (12.1)$$

where ΔH_c is the apparent crystallization enthalpy, w is the weight fraction of isotactic PP in the blends and ΔH_c^0 is the crystallization enthalpy of pure isotactic PP,

The nucleation efficiency can be calculated from dynamic DSC measurements at a constant cooling rate by comparing the crystallization temperatures of the nucleated system with that of the self-nucleated matrix.

The two extremes of the efficiency scale are considered as the non-nucleated and the self-nucleated matrix. Then, the nucleation efficiency, NE, can be considered as a percentage value, and this is given as

$$NE = 100 \frac{T_c - T_{c,0}}{T_{c,max} - T_{c,0}}. \tag{12.2}$$

T_c is the crystallization temperature of the sample under investigation, $T_{c,0}$ is the crystallization temperature of a non-nucleated sample, and $T_{c,max}$ is the crystallization temperature of a best nucleated sample.

Thus the nucleation efficiency is similar to a reduced physical quantity.

12.3 Classes of Nucleating Agents

Adipic acid, benzoic acid, or metal salts of these acids, sorbitols, such as 3,4-dimethylbenzylidene sorbitol are examples of nucleating agents, as are many inorganic fillers (1). Classes of nucleating agents and special examples are given in Table 12.1.

12.3.1 Inorganic Nucleating Agents

For poly(cyanoaryl ether)s, suitable nucleating agents are alumina, aluminum hydroxide, aluminum powder, titanium dioxide and calcium fluoride. In contrast, kaolin, talc, mica, silica, calcium carbonate, a metal salt of an aliphatic acid, etc., which are frequently used for poly(ester)s, exhibit no effective action upon the crystallization of a poly(cyanoaryl ether) (13).

Pyrogenic silicic acid (Aerosil®) has been proposed as a nucleating agent for PA (14). Actually, Aerosil® is a multipurpose agent

Table 12.1: Nucleating Agents (1, 15)

Compound
Talcum
Titanium dioxide
Magnesium oxide
4-*tert*-butylbenzoic acid
Adipic acid
Diphenylacetic acid
Sodium succinate
Sodium benzoate
Ionomers
2,4-Bis-(3,4-dimethylbenzylidene) sorbitol
Disodium[2.2.1]heptane bicyclodicarboxylate
Aluminum 2,2'-methylene-bis(4,6-di-tert-butylphenyl) phosphate
Semi-crystalline branched or coupled poly(olefin)s (9)

and can be used as reinforcement agent, for antiblocking, as well as a rheology improver.

Examples of nucleating agents for poly(esters) are talc, calcium fluoride, sodium phenyl phosphonate, alumina, and in addition, finely divided poly(tetrafluoroethylene).

12.3.2 Sorbitol Compounds

Dibenzylidene sorbitol exists in the form of fibrils and usually acts as an effective nucleating agent to facilitate crystallization of poly-(olefin)s during manufacturing (12). In particular, when the dibenzylidene sorbitol concentration reaches a critical value, the fibrils will self-organize into a three-dimensional network when there is a decrease of temperature, but before crystallization takes place. The network of the fibrils may facilitate the subsequent process of nucleation and crystallization growth. An oriented deformation of the dibenzylidene sorbitol network could act as a template for anisotropic crystallization of PP, which then results in a high lamellar orientation level.

12.3.3 Phosphates

Organic phosphates has been found to have a very high nucleating efficiency achieved even at very low concentrations. This provokes an important increase in the flexural modulus (16).

Organic phosphorous nucleating agents exhibit a better performance on the mechanical properties of i-PP than sorbitol compounds. However, sorbitol compounds more enhance the transparency. On the other hand, organic phosphorous nucleating agents and sorbitol nucleating agents have similar effects on the crystallization peak temperature i-PP (17).

12.3.4 Carbon Nanotubes

Carbon nanotubes can also act as nucleating agents for polymer crystallization.

12.3.5 Coupled Nucleating Agents

Coupled nucleating agents include coupled semi-crystalline homopolymers or copolymers of ethylene, propylene, or other α-olefins. These polymers are coupled together by one or more coupling agents, such as carbenes (18), nitrenes (19), or azide coupling agents (9).

Coupled nucleating agents can be used in the same manner and under the same conditions as conventional nucleating agents. It has been claimed that for PP polymers, the nucleation by coupled or branched polymeric additives is much better than the nucleation by the conventional inorganic nucleators or organic clarifiers.

12.4 Crystallization Accelerators

Crystallization accelerators have been described for PAs. As crystallization accelerators, there are low molecular weight PAs, higher fatty acids, higher fatty acid esters and higher aliphatic alcohols, which may be used independently or in combination.

Table 12.2: Clarifying Agents (4)

Compound
3,4-Dimethylbenzylidene sorbitol (Millad® 3988, Miliken)
p-Methylbenzylidene sorbitol
2,4-Bis-(3,4-dimethyldibenzylidene) sorbitol (Millad® 3940, Miliken)
2,2-Methylene-bis-(4,6-di-*tert*-butylphenyl) phosphate

12.5 Clarifying Agents

The class of clarifying agents overlaps to a wide extent to nucleating agents. In Table 12.2 a few clarifying agents are summarized.

References

1. A.K. Mehta, S. Datta, W. Li, and S.S. Iyer, Heterogeneous polymer blends and molded articles therefrom, US Patent 7 476 710, assigned to ExxonMobil Chemical Patents Inc. (Houston, TX), January 13, 2009.
2. M. Gahleitner and J. Wolfschwenger, "Polymer crystal nucleating agents," in K.H.J. Buschow, R. Cahn, M.C. Flemings, B. Ilschner, E.J. Kramer, S. Mahajan, and P. Veyssiere, eds., *Encyclopedia of Materials: Science and Technology*, pp. 7239–7244. Elsevier, Oxford, 2001.
3. N. Fanegas, M. Gómez, C. Marco, I. Jiménez, and G. Ellis, Influence of a nucleating agent on the crystallization behaviour of isotactic polypropylene and elastomer blends, *Polymer*, 48(18):5324–5331, August 2007.
4. D. Burmaster, O. Hodges, J.L. Lumus, L.A. Kelly, and M. Murphy, Polypropylene having improved clarity and articles prepared therefrom, US Patent 7 241 850, assigned to Fina Technology, Inc. (Houston, TX), July 10, 2007.
5. Standard test method for haze and luminous transmittance of transparent plastics, ASTM Standard, Book of Standards, Vol. 08.01 ASTM D1003-07e1, ASTM International, West Conshohocken, PA, 2009.
6. E. Tobita, K. Nomura, and N. Kawamoto, Nucleating agent composition and crystalline polymer compositions containing the same, US Patent 7 442 735, assigned to Asahi Denka Co., Ltd. (Tokyo, JP), October 28, 2008.
7. Q.F. Yi, X.J. Wen, J.Y. Dong, and C.C. Han, A novel effective way of comprising a [beta]-nucleating agent in isotactic polypropylene (i-pp): Polymerized dispersion and polymer characterization, *Polymer*, 49(23): 5053–5063, October 2008.

8. S. Zhao, Z. Cai, and Z. Xin, A highly active novel β-nucleating agent for isotactic polypropylene, *Polymer*, 49(11):2745–2754, May 2008.
9. J.C. Stevens, D.D. Vanderlende, and P. Ansems, Crystallization of polypropylene using a semi-crystalline, branched or coupled nucleating agent, US Patent 7 250 470, assigned to Dow Global Technologies Inc. (Midland, MI), July 31, 2007.
10. SpecialChem S.A., Nucleation solutions to improve polyolefins performance and processing, [electronic:] http://www.specialchem4polymers.com/tc/Polyolefin-Nucleators/, 2009.
11. M. Valenti, Hyperform® HPN-68L regulatory information document, [electronic:] http://www.hyperformnucleatingagents.com/SiteCollectionDocuments/Hyperform/Documents/HPN-68L%20regulatory%20document.pdf, 2007.
12. K. Wang, C. Zhou, C. Tang, Q. Zhang, R. Du, Q. Fu, and L. Li, Rheologically determined negative influence of increasing nucleating agent content on the crystallization of isotactic polypropylene, *Polymer*, 50(2):696–706, January 2009.
13. S. Matsuo, T. Murakami, T. Bando, and K. Nagatoshi, Reinforced resinous composition comprising polycyano arylene ether, US Patent 4 812 507, assigned to Idemitsu Kosan Company Limited (Tokyo, JP), March 14, 1989.
14. U. Presenz, S. Schmid, R. Hartmann, and H.R. Luck, Method for producing a polyamide compound, US Patent 6 881 477, assigned to EMS-Chemie AG (Domat/Ems, CH), April 19, 2005.
15. P. Piccinelli, M. Vitali, A. Landuzzi, G. Da Roit, P. Carrozza, M. Grob, and N. Lelli, Permanent surface modifiers, US Patent 7 408 077, assigned to Ciba Specialty Chemicals Corp. (Tarrytown, NY), August 5, 2008.
16. C. Marco, M.A. Gómez, G. Ellis, and J.M. Arribas, Highly efficient nucleating additive for isotactic polypropylene studied by differential scanning calorimetry, *Journal of Applied Polymer Science*, 84(9):1669–1679, March 2002.
17. Y.F. Zhang, Comparison of nucleation effects of organic phosphorous and sorbitol derivative nucleating agents in isotactic polypropylene, *Journal of Macromolecular Science, Part B: Physics*, 47(6):1188–1196, 2008.
18. N.C. Mathur, M.S. Snow, K.M. Young, and J.A. Pincock, Substituent effects on the rate of carbene formation by the pyrolysis of rigid aryl substituted diazomethanes, *Tetrahedron*, 41(8):1509–1516, 1985.
19. R.A. Abramovitch, T. Chellathurai, W.D. Holcomb, I.T. McMaster, and D.P. Vanderpool, Intramolecular insertion of arylsulfonylnitrenes into aliphatic side chains, *The Journal of Organic Chemistry*, 42(17):2920–2926, August 1977.

13
Antifogging Additives

13.1 Field of Use

Agricultural films which are largely used in greenhouse culture or tunnel culture chiefly include soft ethylene resin films which are about 20–250 μ thick and which comprise, as a base resin, poly(vinyl chloride), branched low density poly(ethylene) (LDPE), ethylene vinyl acetate (EVA) copolymers, linear low density poly(ethylene) (LLDPE), etc.

Of the various properties required for the agricultural films, particularly important are weather resistance, antifogging properties, heat-retaining properties, and transparency. To cope with the difficult economic situation facing agriculture such as increased costs and a shortage of labor, development of films having an extended duration of life before replacement is desired (1).

The atmosphere in greenhouses or tunnels enveloped in agricultural film is saturated with water vapor that evaporates from the soil or plants, and the water vapor condenses on the inner surface of a cold film to cause fogging. Water droplets on the film not only greatly reduce the incident sunlight due to irregular reflection, but the droplets fall on the plants resulting in frequent occurrence of diseases.

Another issue closely related to the greenhouse problem applies to so called food packaging films when food, e. g. meat products, are packaged on trays and over wrapped with a plastic film at room temperature. When these packages are placed in a refrigerator at around 4°C., the air enclosed in the package cools and is no longer able to hold its water in the vapor phase. The air in the package becomes saturated and the water condenses as water droplets onto

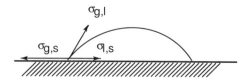

Figure 13.1: Wetting of a Surface

the film's surface.

To overcome these problems, polymer films are modified with antifogging additives. The modified plastic films do not prevent the formation of condensation per se. However, while water vapor condenses on such films, antifogging additives migrate to the surface of the film, causing the condensation to spread evenly over the film's surface and run off instead of forming droplets (2, pp. 609-626).

13.2 Principles of Action

Basically, antifogging agents lower the contact angle between water and the polymer surface. If not complete wetting occurs resulting in the liquid forming a droplet on the surface, as shown in Figure 13.1. If the contact angle becomes sufficiently low, then complete wetting occurs.

13.2.1 Thermodynamics of Surfaces

The work of adhesion W is the work to generate a new surface. The work of adhesion is related to the free energy by the Duprè equation. If a joint is peeled, a certain surface A is lost and but another surface is created.

$$dW = (\sigma_{2,0} + \sigma_{1,0} - \sigma_{1,2})dA. \tag{13.1}$$

$\sigma_{1,0}$ and $\sigma_{2,0}$ are the surface tension against vacuum. The situation changes, if the joint is peeled under water, then $\sigma_{1,W}$ and $\sigma_{2,W}$ must be inserted, which are the surface tensions against water. These constants may significantly differ from the constants for vacuum.

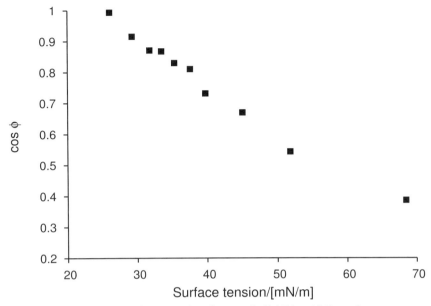

Figure 13.2: Contact Angle at a Solid Liquid Interface

Contact Angle

There is a relationship between the contact angles of certain liquids (L) to solid bodies (S) and adhesives that are chemically related to the liquids.

The contact angle ϕ is related to the surface tension in the equilibrium of forces as

$$\sigma_{S,V} - \sigma_{S,L} = \sigma_{L,V} \cos \phi. \tag{13.2}$$

S, V is the interphase of solid to vacuum, L, V is the interphase of liquid to vacuum, and S, L is the interphase of solid to liquid. Eq. 13.2 is addressed as Young's equation. The situation is illustrated in Figure 13.2. In practice, often it is not possible to establish a vacuum, because of the finite vapor pressure of the liquid. For this reason, in fact, instead of the vacuum, more correctly, saturated vapor or air must be thought.

The critical surface tension of a material can be determined from a Zisman plot (3), which measures the variation in the contact angle as a function of the surface tension of a series of liquids.

Zisman noticed empirically that a plot of $\cos\phi$ versus $\sigma_{L,V}$ is often linear. The value for which $\cos\phi$ extrapolates to 1 is termed as the critical surface tension. From this empirical experience the following equation arises:

$$\cos\phi = 1 - k(\sigma_{L,V} - \sigma_c). \qquad (13.3)$$

Here, $\sigma_{L,V}$ is the surface tension of the liquid, σ_c is the critical surface tension of the solid, i.e., when total wetting will occur.

Zisman himself emphasized that his method is only applicable to pure liquids. He especially warns against using solutions from which different components can be preferentially adsorbed (4).

Several authors have tried to determine critical surface tensions for solid surfaces by determining the contact angles for a set of solutions of different concentrations. Zisman's method is, however, not applicable to solutions due to the large probability for specific and selective adsorption of the components constituting the solution.

A Zisman plot of a wood adhesive with varying concentrations of aqueous acetic acid is shown in Figure 13.3. Comparing the data obtained with the acetic acid solutions to the data obtained with the solutions of the dissolved wood-related substances, considerable discrepancies between the apparent critical surface tension arrived at with the mixed solutions compared with the critical surface tension arrived at with the pure probe liquid have been observed (not shown in Figure 13.3) (4).

13.3 Conventional Compounds

Representative antifogging additives are glycerol monooleate, poly-(glycerol ester)s, sorbitan esters, ethoxylated sorbitan esters, nonylphenol ethoxylate or ethoxylated alcohols (5).

The typical structure of an ethoxylated sorbitan ester is shown in Figure 13.4.

Antifogging additives can be incorporated within the polymer matrix as pure additives or as masterbatches or concentrates. Typical antifogging additive concentrations range between 1 and 3%. The additives have the property of migrating to the surface of the film. In a monolayer film, the antifogging additive migrates in both

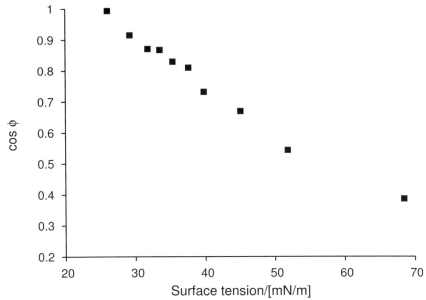

Figure 13.3: Zisman Plot of a Wood adhesive with Varying Concentrations of Aqueous Acetic Acid (4)

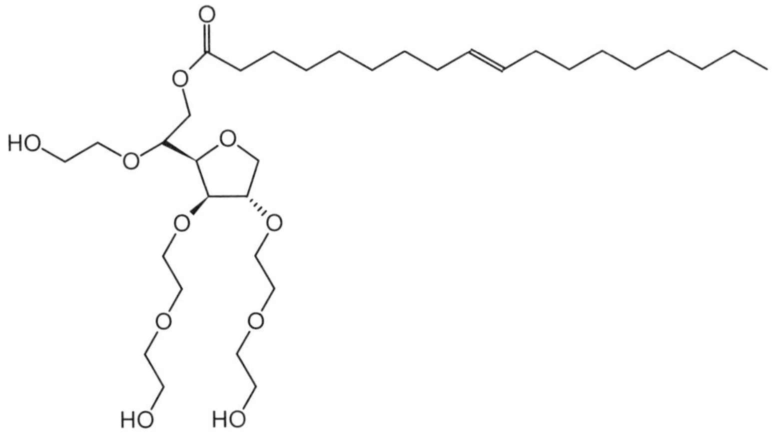

Figure 13.4: Structure of An Ethoxylated Sorbitan Ester

directions, towards the inside of the agricultural film where the antifogging effect is desirable, but also to the outside of the film where it is unnecessary. On the outside of the polymer film, antifogging additive is lost as it is washed off by rain (1).

Antifogging additives can also be applied to the surface by coating. Surfactant molecules coatings have the undesirable property of forming a weak attachment to polymeric films or foils, particularly poly(ethylene) films and are washed away by the action of heat and humidity.

However, an antifogging film obtained by coating a soft plastic film with an antifogging agent has not yet been employed extensively as an agricultural film. Because of their low surface energy, soft plastic films for agricultural use generally have poor wettability and adhesion when coated with surface active agents or hydrophilic high polymeric substances when used as antifogging agents. This tendency is particularly conspicuous with soft ethylene resin films of low polarity, e.g., LDPE, EVA, and LLDPE films. Therefore, when an antifogging agent is spray coated with a power atomizer onto a soft ethylene resin film, the antifogging agent needs to be used in a large quantity which increases costs, because a lot of time is required for the spray coating operation. Further, spray coating cannot be effected uniformly so that the antifogging effects are not uniform either.

Where an antifogging agent is applied using a coater, etc., a large quantity of a coating is consumed, and the coating speed cannot be increased, resulting in an increase of costs. In either case, the coated antifogging agent is washed away with running water droplets due to poor adhesion resulting in a very short life for the antifogging properties. Furthermore, the coated film undergoes blocking due to the stickiness of the antifogging agent. As a result, it has been impossible to retain antifogging effects in a stable manner for a long duration of at least one year, more desirably, several years. Most of the state-of-the-art agricultural films exhibit antifogging properties from coated additives for a period of only about one month.

Antifogging agents commonly incorporated into the films include non-ionic, anionic and cationic surface active agents. Other methods for providing antifogging properties to agricultural films, in addition to the coating method and incorporation method, include chemical modification of the ethylene base resin or the ethylene resin film surface by introducing a polar group, such as a hydrophilic group. This technique, however, entails high cost at the present time and is difficult to apply.

Suitable inorganic hydrophilic colloidal substances include colloidal silica, colloidal alumina, colloidal $Fe(OH)_2$, colloidal $Sn(OH)_4$, colloidal TiO_2, colloidal $BaSO_4$, and colloidal lithium silicate, with colloidal silica and colloidal alumina most generally used. Suitable hydrophilic organic compounds include (1):

- Non-ionic, anionic or cationic surface active agents
- Graft copolymers with mainly a vinyl monomer unit that contains a hydroxyl moiety and from 0.1 to 40% by weight of a carboxyl-containing vinyl monomer unit
- Sulphur-containing poly(ester) resins.

13.4 Compounds for Grafting

One way to obtain permanent additives is to covalently bond them to the polymeric matrix through a chemical reaction. Among possible reactions, one option is to have photo-reactive moieties in the additive molecule so that, by exposition to visible and/or ultraviolet light, either from an artificial source or from solar irradiation, the molecule reacts with the polymeric substrate. As a consequence, the additive are grafted to the polymer so that the effect imparted by the former is permanent. Several examples of photo-reactive additives are reported in the literature (6).

In Figure 13.5 a compound is shown that can be fixed onto a polymeric matrix.

This compound is an example of a graftable functional monomer having in addition a hydrophilic residue. Composition may undergo photo-grafting via exposition under natural light or artificial suitable UV source. Responsible for photo-grafting is the cinnamic moiety in the compound.

By combining opportunely migration of antifogging agents to the surface of the polymer substrate with the reaction of a photo-sensible moiety contained in the antifogging additive itself, it is possible to have the reaction induced by light when most or at least part of the

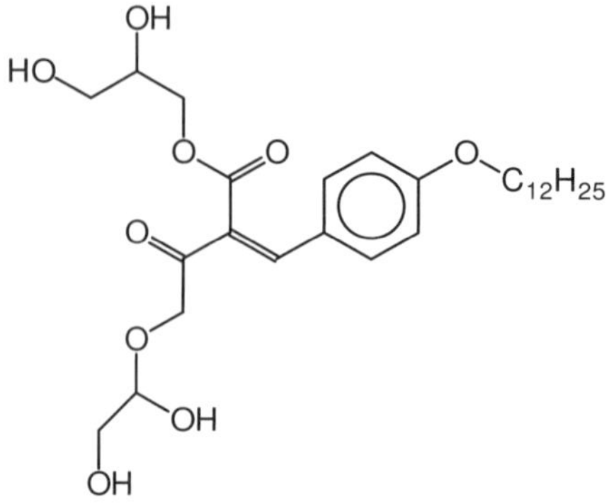

Figure 13.5: 2-(4-Dodecyloxy-benzylidene)-malonic acid bis-(2,3-dihydroxy-propyl) ester (1)

additive is at the surface of the plastic film. This results in long-term antifogging properties because the additive is permanently bound to the polymer and cannot be physically removed (1).

Another option is to apply the antifogging additive to the surface by coating (e.g., by spraying or by roller-coating) and, after that, induce the reaction between the photo-graftable part of the additive and the polymer either by a proper artificial treatment or by directly exposing the plastic film to the natural light. In this case the reaction must be fast enough, in order to occur before the additive is washed off by humidity. Similar to the previous example, the antifogging effect is then retained for a long time (1).

Of interest are also properties other than fogging resistance. Water repellency and oil repellency in particular are the relevant properties in this context. The former can reduce dust deposition, through mechanical removal of water rapidly flowing along the repellent polymer surface and may find application in greenhouse films for agriculture application, in order to enhance light transmittance inside the greenhouse. Oil repellency has the consequence to impart stain resistance to fabrics made of fibers or nonwovens.

References

1. P. Piccinelli, M. Vitali, A. Landuzzi, G. Da Roit, P. Carrozza, M. Grob, and N. Lelli, Permanent surface modifiers, US Patent 7 408 077, assigned to Ciba Specialty Chemicals Corp. (Tarrytown, NY), August 5, 2008.
2. H. Zweifel, ed., *Plastics Additives Handbook*, Hanser Publishers, Munich, 5th edition, 2001.
3. W.A. Zisman, "Contact angle, wettability and adhesion," in F.M. Fowkes, ed., *Kendall Award Symposium*, Vol. 43 of *Advances in Chemistry*, pp. 1–51. American Chemical Society, Washington, DC, 1964.
4. J. Nylund, K. Sundberg, Q. Shen, and J.B. Rosenholm, Determination of surface energy and wettability of wood resins, *Colloids and Surfaces A: Physicochemical and Engineering Aspects*, 133(3):261–268, February 1998.
5. K. Sudo, S. Mori, T. Okada, H. Ikeno, K. Shichijo, and S. Ohnishi, Agricultural film, US Patent 5 262 233, assigned to Mitsubishi Petrochemical Co., Ltd. (Tokyo, JP), November 16, 1993.
6. M. Mehrer, T. Staehrfeldt, and M. Zaeh, Novel 2-(2'-hydroxyphenyl)-benzotriazoles and 2-hydroxybenzophenones as light protectors for polymeric materials, EP Patent 0 897 916, assigned to Clariant GmbH, February 24, 1999.

14
Antiblocking Additives

Blocking is the unwanted adhesion between layers of plastic film that may occur under pressure, usually during storage or use (1). Due to blocking it is difficult to unwind a film roll or to open a bag. Blocking can be prevented with the use of antiblocking agents that are added to the composition which makes-up the surface layer of the film (1).

If inorganic mineral antiblocking agents, such as talc or silica are dispersed throughout the film, the film surface is roughened on a microscopic level. For this reason, adjacent film layers will not stick. Also, organic antiblocking agents are in use, including fatty amides, stearates, silicones, and poly(tetrafluoroethylene) (2).

In films, the optical properties, i.e., clarity or transparency, respectively, may be a major concern. Clarity is directly related to the difference in the refractive index of the additive and the polymeric matrix. If the refractive index of the antiblocking agent matches the refractive index of the polymer, the optical properties are left unchanged. Further, clarity depends on the particle size distribution and the amount of additive used. Larger antiblocking particles exhibit a higher blocking effect, although they basically negatively influence the optical properties (2). To achieve high clarity organic amides should be used alone. However, organic materials may migrate eventually to the surface.

Other concerns may be interactions with other additives. For example, antiblocking agents may absorb other additives such as antioxidants, slip agents, and other processing aids. Natural silica shows a low level of interaction whereas synthetic silica types and uncoated talc exhibit a higher absorption of additives. These properties may be affected by surface treatment (2).

Table 14.1: Antiblocking Agents (3)

Compound
Silica
Clay
Limestone
Talc
Zeolithes
Glass
Synthetic waxes

Further, abrasion may be a problem in the course of processing if the hardness of the inorganic antiblocking agent is too high. For example, talc is much more soft than silica and causes a least abrasive level to the processing engines (2).

14.1 Examples of Uses

14.1.1 Film Resins

Antiblocking agents for film resins are summarized in Table 14.1. They find application for linear low density poly(ethylene), poly(propylene), poly(vinyl chloride), and poly(ethylene terephthalate) in amounts of 1000–4000 ppm. These antiblocking additives are also used for a heat-sealable outer skin of a multilayer film (3).

The possibility of using silica from rice husk ash as an antiblocking agent in low density poly(ethylene) films has been investigated. Silica from rice husk ash has a smaller particle size and thus a higher specific surface area in comparison to other commercially available silica types (4).

Unmodified multilayer film with heat-sealable skin layers has an inherently high coefficient of friction and film-to-film blocking properties. Therefore, slip additives and antiblocking particulates are traditionally added to the film structure to lower the coefficient of friction and provide improved machinability to produce, for example, food packages.

The slip properties of multilayer film have been beneficially modified by the inclusion of polymers of fatty acid amides. Fatty acid amide materials, however, disadvantageously depend on film temperature and storage time to promote the migration and effectiveness of this type of slip system. Fatty acid amide slip systems also have reduced functionality when the film is laminated to other non-slip containing films. Namely, the coefficient of friction increases after lamination. Therefore, the production and functionality of fatty acid amide slip systems is limited (5). However, an improved coefficient of friction and slip functionality can be gained by the incorporation of silicone oil into the skin layer of a multilayer film.

A disadvantage with silicone oil slip systems, however, is the difficulty in converting a multilayer film that employs a silicone oil slip system. Silicone oil tends to transfer from one film surface to another upon winding of the film. Due to the silicone oil lubrication on both sides of the film, the treated surface becomes contaminated and consequently makes printing and ink adhesion more difficult. Additionally, if printing and laminating are done in two steps, i.e., out-of-line, then silicone oil can also transfer to the surface of the ink and cause future lamination bonding strengths to be low or inconsistent (5). Compared to silicone oil, silicone gum tends to migrate less throughout the multilayer film and tends to transfer less from one film surface to another upon winding of the film.

In conclusion, the disadvantages can be reduced by a co-extruded, heat-sealable film structure, comprising (5):

- A core layer of a thermoplastic polymer, the core layer having a first side and a second side
- A functional layer on the first side of the core layer, wherein the functional layer is a laminating layer, a printable layer, a laminating and a printable layer, or a sealable layer
- A heat-sealable layer on the second side of the core layer comprising a thermoplastic polymer and an amount of a slip system, based upon the entire weight of the heat-sealable layer, sufficient to reduce the coefficient of friction and improve slip performance of the heat-sealable layer, wherein the slip system comprises silicone gum and antiblocking agents.

Surging

Films with improved slip characteristics are obtained by adding 1000 ppm polydimethylsiloxane, together with 1000 ppm erucamide, to a poly(olefin) (6). However, such a formulation cannot be extruded from a single screw extruder, in a satisfactory manner, because of surging.

During extrusion, the composition oscillates from periods of slipping, to periods in which slipping does not occur. Due to surging, the film exhibits undesirable fluctuations in thickness and silicone oil concentration.

Surging does not occur upon the extrusion of a composition comprising propylene/ethylene random copolymer and additionally aluminum silicate. The film produced from such a formulation is free from undesirable fluctuations in thickness and organosiloxane concentration (6).

A relatively high level of aluminum silicate antiblocking agent prevents the presence of the relatively high level of organosiloxane from causing the intermittent screw slippage resulting in surging. Thus, it is apparent for the production of high quality heat shrinkable packaging films, the silicone oil and antiblocking agent must act together to provide a plurality of advantages, each allowing the other to be used, to achieve a film having the desired coefficients of friction, and a uniformity of thickness and organosiloxane concentration.

Without the presence of the relatively high level of silicone oil, the relatively high level of the antiblocking agent would result in an undesirable head pressure in the extruder, thereby lowering the extrusion speed, as well as resulting in a higher melt temperature and a lower output rate from the extruder (6).

14.1.2 Sealable Coatings

Sealable coatings are used on flexible packaging films so that the films can be sealed with the application of pressure, with or without out exposure to elevated temperatures. These so called cold seal coatings can pose blocking problems. A typical cold seal coating is a natural or synthetic rubber latex combined with a soft polymer which tends to be tacky at room temperature and causes blocking. The rubber component permits sealing with slight pressure and without using heat. The cold seal coating is usually applied to a plastic film as it is wound into a roll. Since the cold seal coatings are tacky, it is important that the backside of the film which contacts the cold seal coating upon winding does not stick (block) to the cold seal coating so that the film can be easily unwound for use on packaging equipment (1).

One approach for reduced blocking between the cold seal coating and the backside of the film has been to formulate a cold seal coating which is nonblocking to certain surfaces including poly(propylene), such a cold seal formulation is described in U.S. Pat. No. 5,616,400.

Another approach uses a cold seal release material on the layer opposite the cold seal surface (7).

A film is described which has an upper heat-sealable layer formed from an ethylene-propylene containing copolymer and an antiblocking agent. The lower heat-sealable layer is formed from an ethylene-propylene containing copolymer and antiblocking agent and silicone that reduces friction. The silicone oil additive has a viscosity of 10,000–30,000 cSt. The silicone is present on the exposed surface discrete microglobules. These microglobules transfer to the upper surface upon contact. The silicone on the surfaces of the film facilitates the machinability (3).

An attempt was made to produce a block-resistant functional film, typically a film having a printing function or sealing function, with silicone oil in a surface layer as an antiblocking agent. It was found that the silicone oil was detrimental to the printing or sealing function.

Likewise block-resistant films are composed from poly(dialkylsiloxane). When the film is wound into a roll, poly(dialkylsiloxane) deposits silicone onto the functional layer but the amount of silicone deposited is not substantially detrimental to the printing function or the sealing function (1).

14.1.3 Membranes

A breathable poly(urethane) membrane has been manufactured by using a multilayer film extrusion technique. In the course of this process, the membrane is blown as the inner layer of a three-layer film. The outer support layers are based on poly(ethylene). It was found that the neat membrane surface was unacceptably sticky. However, starch is a suitable antiblocking agent and better than mineral fillers. The use of starch decreases the permeability of the membrane towards moisture. In addition, theoretical studies confirm the impermeable nature of the starch particles (8). This rather unexpected finding can be explained by assuming a crystalline or glassy structure of the starch domains.

14.1.4 Poly(vinyl butyral)

Plasticized poly(vinyl butyral), whether in sheet or pellet form, inherently tends to stick to itself during storage or transportation. The strength of the self-adhesion of plasticized poly(vinyl butyral) can reach such levels that it is impossible to separate if the sheets or pellets were not refrigerated. This kind of severe self-adhesion is referred to as blocking. The nature of blocking creates great difficulties in the sheet handling during manufacture, transportation, and the lamination process. It also makes it extremely difficult to feed poly(vinyl butyral) pellets continuously into an extruder (9).

Fatty acid amides are well known as an important class of polymer additives used as a slip agent or lubricant to prevent unwanted adhesion. Unfortunately the addition of such amides for the purpose of antiblock additives for polymer sheets, in glass laminates (reduction of unwanted adhesion between surfaces of the polymer sheet itself) adversely affects the optical characteristics of the polymer sheet, such as haze, transparency, and film clarity, as well as adhesion of the polymer sheet to glass. Consequently, the inclusion of fatty acid amides polymer sheet interlayers in glass laminates is necessary.

Certain fatty acid amides, however, can be successfully used as an antiblocking agent in polymer sheets while not affecting optical properties of the polymer sheet or the adhesive properties of the polymer sheet to glass. These amides include erucamide, behenamide (docosanamide), *N*-oleyl palmitamide, stearyl erucamide, erucyl stearamide, hydroxystearamide, oleic acid diethanolamide, stearic acid diethanolamide, poly(ethylene glycol) oleic amide, and mixtures of these amides. Secondary mono-amides are particularly preferred. A particularly preferred secondary mono-amide is *N*-oleyl palmitamide, an amide with a double bond geometry (9). Some organic antiblocking agents are shown in Figure 14.1.

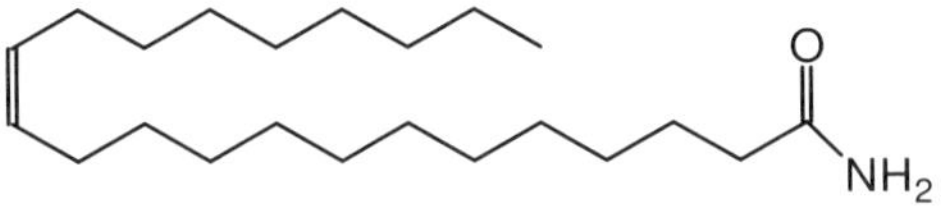

Erucamide

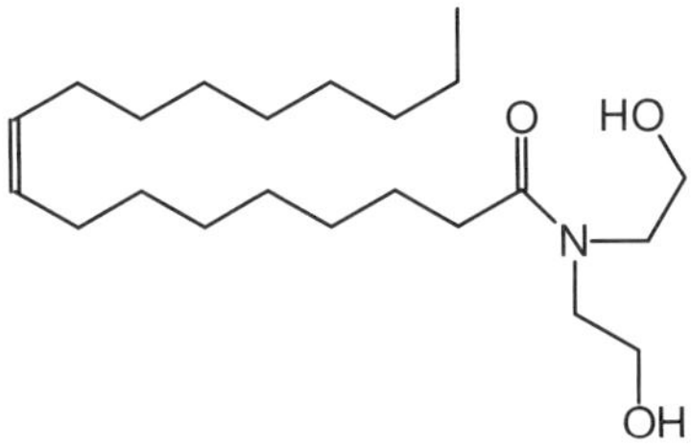

Oleic acid diethanolamide

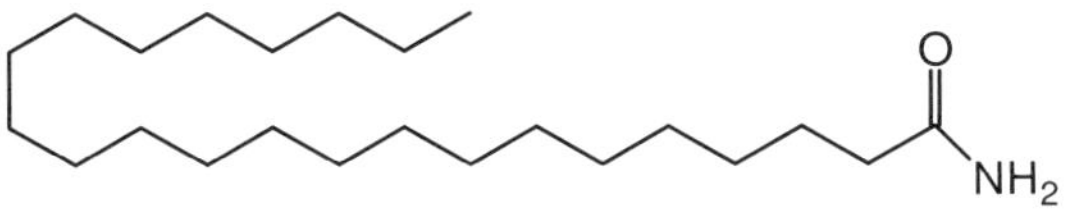

Behenamide

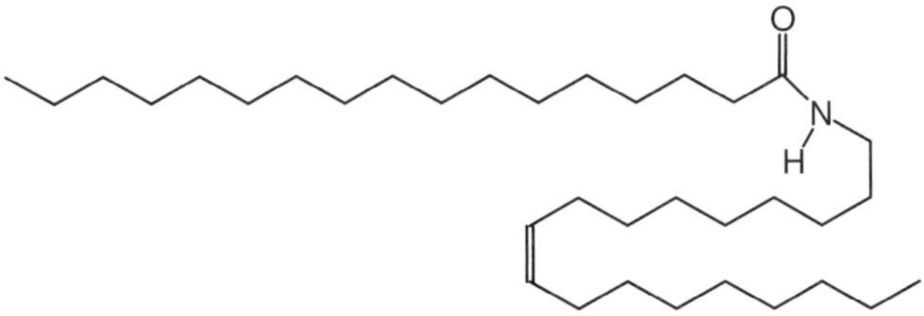

N-Oleyl palmitamide

Figure 14.1: Organic Antiblocking Agents

References

1. G.F. Cretekos and J.R. Wagner, Jr., Block-resistant film, US Patent 6 472 077, assigned to Exxon Mobil Oil Corporation (Fairfax, VA), October 29, 2002.
2. J. Markarian, Slip and antiblock additives: Surface medication for film and sheet, *Plastics, Additives and Compounding*, 9(6):32–35, November–December 2007.
3. J.K. Keung, K.M. Donovan, and R. Balloni, Heat sealable film and method for its preparation, US Patent 4 692 379, assigned to Mobil Oil Corporation (New York, NY), September 8, 1987.
4. S. Chuayjuljit, C. Kunsawat, and P. Potiyaraj, Use of silica from rice husk ash as an antiblocking agent in low-density polyethylene film, *Journal of Applied Polymer Science*, 88(3), 2003.
5. K.A. Sheppard and R.A. Migliorini, Lamination grade coextruded heat-sealable film, US Patent 7 393 592, assigned to ExxonMobil Oil Corporation (Irving, TX), July 1, 2008.
6. G.P. Shah and G.J. Hayes, Film containing silicon oil and antiblocking agent, US Patent 6 291 063, assigned to Cryovac, Inc. (Duncan, SC), September 18, 2001.
7. A.F. Wilkie, Biaxially and monoaxially oriented polypropylene cold seal release film, US Patent 5 489 473, assigned to Borden, Inc. (Columbus, OH), February 6, 1996.
8. S. Pecku, T.L. van der Merwe, H. Rolfes, and W.W. Focke, Starch as antiblocking agent in breathable polyurethane membranes, *Journal of Vinyl and Additive Technology*, 13(4), 2007.
9. W. Chen, A.N. Smith, and A. Karagiannis, Poly(vinyl butyral) pellets, US Patent 7 491 761, assigned to Solutia Incorporated (St. Louis, MO), February 17, 2009.

15
Hydrolysis

Not all polymers are prone to hydrolytic degradation. For example the backbone of poly(olefin)s is essentially inert to hydrolytic degradation. However, poly(ester)s and poly(amide)s are more or less sensitive to hydrolytic degradation.

15.1 Hydrolytic Degradation

I discuss below the variants of hydrolytic degradation. These variants include ordinary hydrolysis and enzymatic hydrolysis.

15.1.1 Ordinary Hydrolysis

In particular, polymers with ester or amide linkages are prone to hydrolytic degradation. Basically the hydrolysis for an ester proceeds as

$$CH_3-COO-CH_2-CH_3 + H_2O \rightarrow CH_3-COOH + HO-CH_2-CH_3. \tag{15.1}$$

Thus, from an ester, an acid and an alcohol is formed. Since the reaction is catalyzed by acids, the acid formed in the course of the reaction acts as a catalyst. This type of reaction is termed *autocatalytic*. In the same manner, amide links decompose into acids and amines,

$$CH_3-CONH-CH_2-CH_3 + H_2O \rightarrow CH_3-COOH + H_2N-CH_2-CH_3. \tag{15.2}$$

The hydrolysis reactions are just the reverse reactions of the respective condensation reactions.

15.1.2 Enzymatic Hydrolysis

In biodegradable plastics, e.g., polyesters or poly(lactide)s, special monomers are used that can easily be metabolized by microorganisms. In general, in a biodegradable plastic, biodegradation proceeds by the following processes (1):

1. A polymer decomposition enzyme adsorbs on the surface of a polymer material. The enzyme is one such as extracellularly secreted by a specific kind of microorganism.
2. The enzyme cleaves chemical bonds such as ester, glycoside and peptide linkages in polymer chains by hydrolysis reaction.
3. The polymer material is further decomposed up to a monomer unit level by the decomposition enzyme with decrease in molecular weight.
4. Eventually, the decomposed products are further metabolized and consumed to be converted to carbon dioxide, water and cell components.

15.1.3 Stabilization

Carbodiimides are preferred as the hydrolytic stabilizers. The basic mechanisms have been been discussed and are summarized in Figure 15.1. Polymeric carbodiimides are superior, because of their multiple functionality that can act as crosslinkers.

A poly(carbodiimide) can be synthesized by heating an organic diisocyanate in the presence of a carbodiimidation catalyst (1, 2). Cyclic phosphine oxides, such as 3-methyl-1-phenyl-3-phosphorene-1-oxide are suitable catalysts.

15.2 Polymers

As already indicated, several classes of polymers are prone to hydrolysis reactions. These polymers bear hydrolyzable groups mostly in

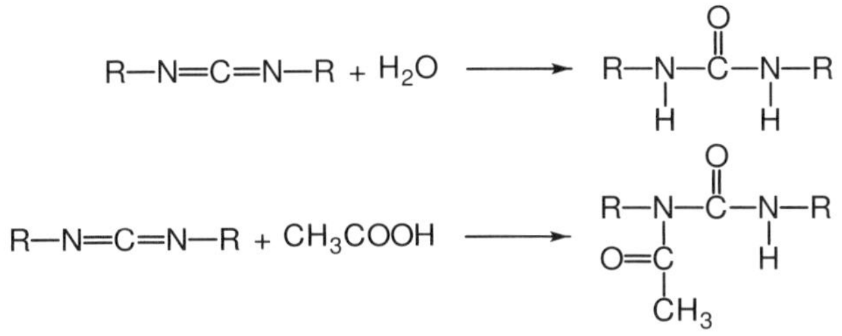

$$R-N=C=N-R + H_2O \longrightarrow \underset{H}{R-N}-\underset{}{\overset{\overset{O}{\|}}{C}}-\underset{H}{N-R}$$

$$R-N=C=N-R + CH_3COOH \longrightarrow \underset{\underset{CH_3}{\overset{O=C}{|}}}{R-N}-\overset{\overset{O}{\|}}{C}-\underset{H}{N-R}$$

Figure 15.1: Stabilization of Hydrolytic Degradation (3)

their main chains.

15.2.1 Poly(ester)s

Linear and crosslinked poly(ester)s are subject to hydrolytic aging. The hydrolysis of the ester linkage leads to a chain scission which results in brittle materials.

In crosslinked unsaturated poly(ester)s, the crosslinking density can serve in evaluating the extent of degradation. However, in the range of high conversions, spectrometric measurements of the ester concentration are recommended (4).

In linear poly(ester)s, predictions of the end of service time can be based on the relationship between the ductile-brittle transition and the entanglement limit of the molecular weight. However, for crosslinked poly(ester)s, this method is not applicable. The reasons for complications in the degradation behavior arise from the diffusion control of hydrolysis kinetics, heterogeneity of semi-crystalline polymers, and variation of the hydrophilicity with the hydrolysis conversion.

15.2.2 Poly(ester urethane)s

Since poly(ester) poly(urethane)s bear ester groups they are likewise subject to hydrolytic degradation (5).

It has been shown that hydrolysis can play a significant role in the aging of these materials (6). Moreover, the ester segments are much more susceptible to hydrolytic a cleavage than the urethane

(7). The molecular weight changes resulting from hydrolysis, have been shown to have a large impact on the physical and mechanical properties of the elastomer (8).

In addition, poly(urethane)s and segmented poly(urethane urea)s derived from lysine diisocyanate, can be enzymatically degraded by various proteases. Thiol proteases, such as papain, bromelain, and ficin, show a high activity. In addition, protease K and chymotrypsin hydrolyze the poly(urethane)s (9).

Hydrolytic stabilizers can be included in poly(ester) poly(ol) based poly(urethane)s. Two commercially available carbodiimide based hydrolytic stabilizers are known as Stabaxol® P and Stabaxol® P-100. They are available from Rhein Chemie of Trenton, N.J., and are effective at reducing the susceptibility of the material to hydrolysis (5).

Still other hydrolytic stabilizers, such as those which are carbodiimide or poly(carbodiimide) based, or based on epoxidized soy bean oil, are considered useful. The total amount of hydrolytic stabilizer employed will generally be less than 5.0% of the composition (10).

Besides of hydrolytic stabilizers, the hydrolytic stability of poly(urethane)s can be tailored by the structure of the backbone.

For example, a polyester polyol synthesized from 2,4-diethyl-1,5-pentanediol exhibits a superior hydrolytic resistance (11).

15.2.3 Poly(lactide)s

Lactic acid based polymers manufactured by melt processing are blow molded bottles, injection molded cups, spoons, and forks. Other applications include paper coatings, fibers, films, and various medical devices.

The hydrolysis of the polymers leads to molecular fragmentation. The hydrolytic degradation of lactic acid based polymers is an undesired phenomenon under certain circumstances, such as during processing or material storage. It is however, beneficial in other applications, for example in medical devices or compostable packages (12).

In branched poly(lactide)s, the enzymatic degradability and the hydrolytic degradation increase with the branches present in the molecule (13).

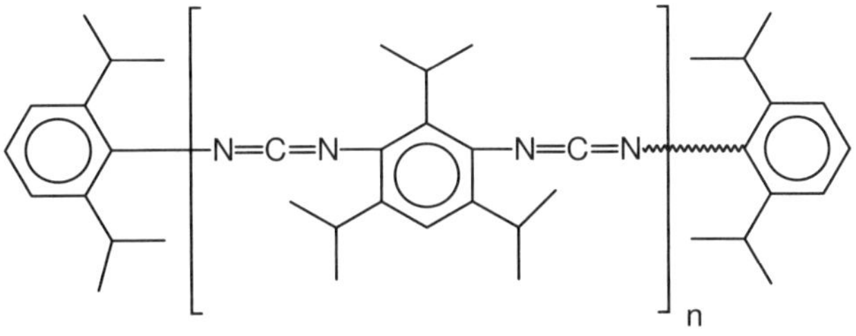

Figure 15.2: Poly(carbodiimide): Stabaxol® P (3)

Poly(carbodiimide)s have been used to improve the thermal stability poly(lactic acid) (PLA) during processing. An addition of the poly(carbodiimide) in an amount of 0.1–0.7% could stabilize the PLA at 210°C up to 30 min (3). The Poly(carbodiimide) is shown in Figure 15.2.

In conclusion, I have shown that the mechanism of stabilization has been investigated by allowing to react the carbodiimide with various model substances, e.g., water, acetic acid, ethanol. The structure of the products of solvolysis has also been elucidated. It has also been demonstrated that the carbodiimide reacts with residual or newly formed moisture and lactic acid, or the carboxyl and hydroxyl end groups in the PLA. These groups are hampering the thermal degradation and hydrolysis of PLA.

References

1. I. Takahashi, H. Iida, and N. Nakamura, Carbodiimide composition with suppressed yellowing, a stabilizer against hydrolysis and a thermoplastic resin composition, US Patent 7 368 493, assigned to Nisshinbo Industries, Inc. (Tokyo, JP), May 6, 2008.
2. Y. Takiguchi, K. Yahata, Y. Komoto, A. Hayashida, and M. Takamizawa, Process for the preparation of a polycarbodiimide solution, US Patent 5 750 637, assigned to Shin-Etsu Chemical Co., Ltd. (Tokyo, JP), May 12, 1998.
3. L. Yang, X. Chen, and X. Jing, Stabilization of poly(lactic acid) by polycarbodiimide, *Polymer Degradation and Stability*, 93(10):1923–1929, October 2008.

4. V. Bellenger, M. Ganem, B. Mortaigne, and J. Verdu, Lifetime prediction in the hydrolytic ageing of polyesters, *Polymer Degradation and Stability*, 49(1):91–97, 1995.
5. G. Oertel, W. Becker, and D. Braun, eds., *Polyurethane*, Vol. 7 of *Kunststoff Handbuch*, Carl Hanser Verlag, Munich, 2nd edition, 1983.
6. C.S. Schollenberger and F.D. Stewart, Thermoplastic polyurethane hydrolysis stability, *Journal of Elastomers and Plastics*, 3(1):28–56, 1971.
7. L. Fambri, A. Penati, and J. Kolarik, Synthesis and hydrolytic stability of model poly(ester urethane ureas), *Angewandte Makromolekulare Chemie*, 228(1):201–219, 1995.
8. D.G. Thompson, J.C. Osborn, E.M. Kober, and J.R. Schoonover, Effects of hydrolysis-induced molecular weight changes on the phase separation of a polyester polyurethane, *Polymer Degradation and Stability*, 91(12):3360–3370, December 2006.
9. N. Yamamoto, A. Nakayama, M. Oshima, N. Kawasaki, and S.i. Aiba, Enzymatic hydrolysis of lysine diisocyanate based polyurethanes and segmented polyurethane ureas by various proteases, *Reactive and Functional Polymers*, 67(11):1338–1345, November 2007.
10. H.W. Bonk and D.J. Goldwasser, Membranes of polyurethane based materials including polyester polyols, US Patent 7 078 091, assigned to Nike, Inc. (Beaverton, OR), July 18, 2006.
11. S. Murata, T. Nakajima, N. Tsuzaki, M. Yasuda, and T. Kato, Synthesis and hydrolysis resistance of polyurethane derived from 2,4-diethyl-1,5-pentanediol, *Polymer Degradation and Stability*, 61(3):527–534, 1998.
12. A. Sodergård and M. Stolt, Properties of lactic acid based polymers and their correlation with composition, *Progress in Polymer Science*, 27(6):1123–1163, July 2002.
13. K. Numata, R.K. Srivastava, A. Finne-Wistrand, A.C. Albertsson, Y. Doi, and H. Abe, Branched poly(lactide) synthesized by enzymatic polymerization: Effects of molecular branches and stereochernistry on enzymatic degradation and alkaline hydrolysis, *Biomacromolecules*, 8(10):3115–3125, October 2007.

16

Dehydrochlorination Stabilizers

Dehydrochlorination stabilizers serve mainly as an additive for poly(vinyl chloride) (PVC). PVC is sensitive to dehydrochlorination.

16.1 Dehydrochlorination of PVC

When PVC is processed at high temperatures, it degrades mainly by dehydrochlorination. Regular sequences degrade in a different way from irregular sequences. This issue is shown in Figure 16.1.

The presence of irregular sequences increases the rate of degradation considerably. The initial rates of degradation are proportional to the content of these irregularities.

The degradation is autocatalytic, i.e., the hydrogen chloride formed initially, catalyzes the reaction further. Moreover, the degradation of PVC proceeds very fast in the presence of Lewis acids, such as $FeCl_3$, $ZnCl_2$, $AlCl_3$, etc.

It has been proposed that the Lewis acids form intermediate complexes like

When the dehydrochlorination reaction proceeds, a sequence of conjugated double bonds are formed that causes coloring.

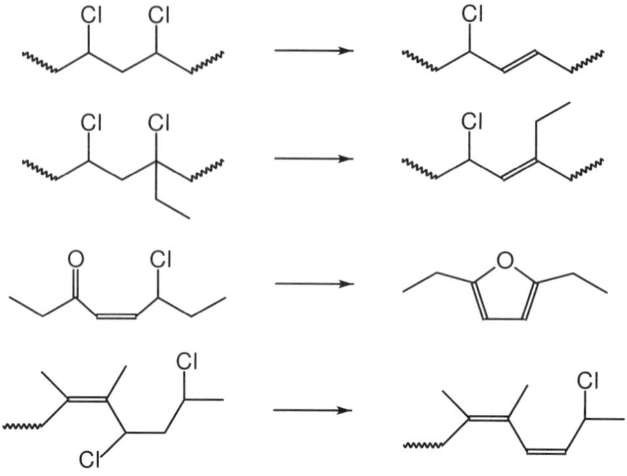

152 Additives for Thermoplastics

Figure 16.1: Degradation of PVC. Top: Regular Sequence. Below: Some Irregular Sequences

Table 16.1: Types of Stabilizers

Substance Class
Alkyl tin compounds
Mixed metal compounds
β-Diketones
Epoxidized fatty acids
Hydrotalcite compounds

In presence of oxygen, the allylic position is sensitive to the attack by oxygen. The primary reaction is shown in Figure 16.2.

In the course of secondary reactions, crosslinking can occur by a Diels-Alder reaction. This reaction is shown in Figure 16.3.

16.2 Stabilizers

In Table 16.1, the types of stabilizers for PVC are listed.

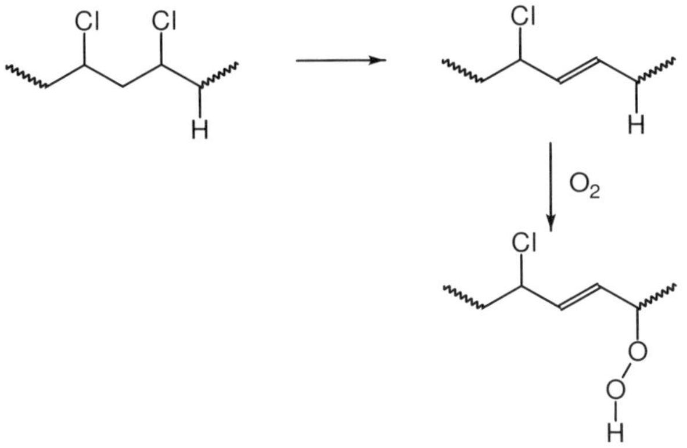

Figure 16.2: Thermooxidative Degradation of PVC

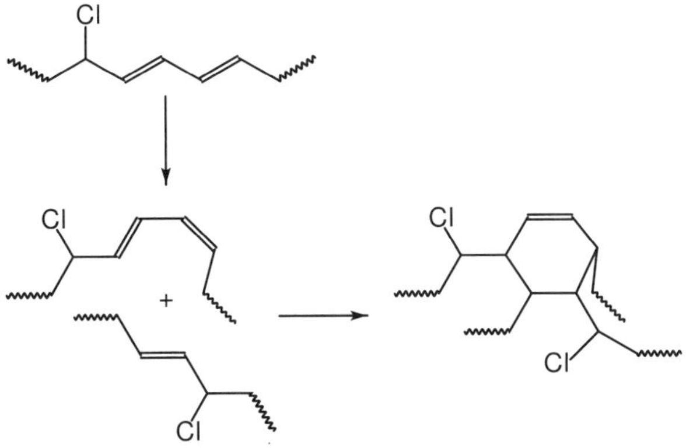

Figure 16.3: Diels-Alder Crosslinking

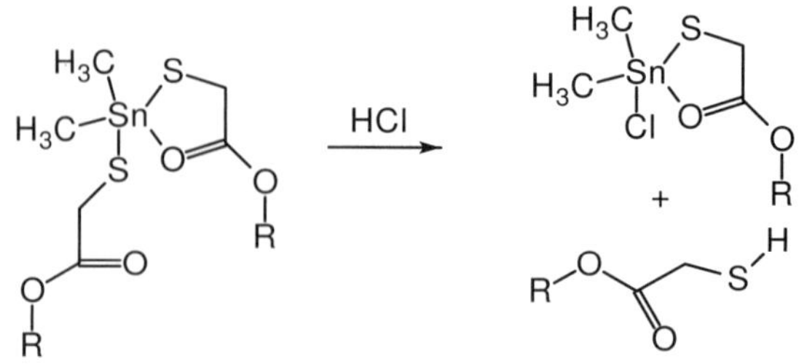

Figure 16.4: Alkyl Tin Stabilizers

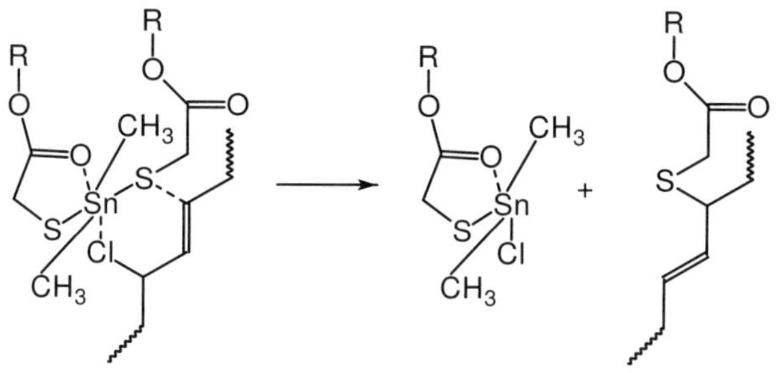

Figure 16.5: Mechanism of Stabilization by Thioglycolate Units

16.2.1 Alkyl Tin Compounds

The most important stabilizers for PVC are alkyl tin stabilizers, e.g., octyltin thiogylcolate. The mechanism of reaction is shown in Figure 16.4.

A thiogylcolate is set free when HCl evolves from the degradation of PVC. Thus, these stabilizers are addressed as secondary stabilizers. The thiogylcolate with its mercaptan moiety, fixes on the PVC backbone to prevent further hydrogen chloride ejections.

In particular, the mercapto groups attack chlorine atoms that are in an allylic position to previously formed double bonds. These chlorine atoms are substituted by the sulfur moiety, which is attached to a bulky rest.

16.2.2 Mixed Metal Compounds

Mixed metal stabilizers are the salts of long chain carboxylic acids, also addressed as metal soaps. Actually, the hydrogen chloride evolved in the course of degradation of PVC is scavenged by the addition of metal soaps in the same sense as a buffer substance acts in ordinary acid base chemistry. However, the action is more complicated, as the affectivity is dependent on the nature of the metal ion. Moreover, synergistic effects using tin or cadmium, together with barium or calcium salts, have been observed. This behavior is attributed to the formation of complexes.

Compounds of the heavy metals lead, barium, and cadmium are particularly well suited as stabilizers, but are nowadays subject to environmental concerns due to their heavy metal content (1).

Long chain carboxylates are steareates, palmitates, and oleates.

16.2.3 β-Diketones

β-Diketones have a structure like

$$\text{\small{wwv}}\overset{\text{O}}{\underset{\|}{C}}-CH_2-\overset{\text{O}}{\underset{\|}{C}}\text{\small{wwv}}$$

and they react with tin compounds as catalysts in alkylation reactions. The activity increases with the acidity of the hydrogens of the methylene groups H–C–H between the carbonyl groups.

16.2.4 Epoxidized fatty acids

Epoxidized fatty acids act as scavengers for HCl with the addition of hydrogen chloride. The mechanism is shown in Figure 16.6.

Mixtures composed of a alkanolamine or a reaction product of an epoxide and of an amine and an uracil compound are particularly suitable for stabilizing PVC. Further, the addition of a perchlorate salt, improves the stabilization of PVC. Uracil is also addressed as 2,4-pyrimidindion.

It is naturally occurring as one of the four base pairs in ribonucleosides. The uracil structure is shown in Figure 16.7.

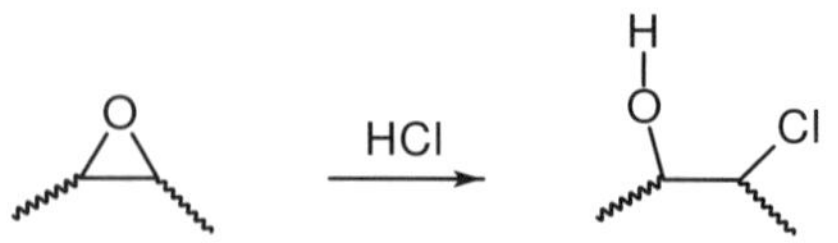

Figure 16.6: Addition of Hydrogen Chloride by Epoxides

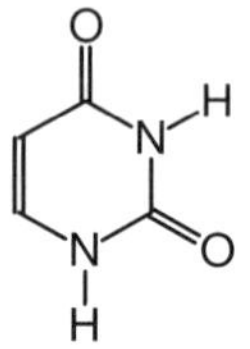

Figure 16.7: Uracil

The use of other classes of amines alone without perchlorate does not give satisfactory processing stability, and in particular the initial color of the desired moldings does not differ substantially from that of an unstabilized specimen (1).

β-Hydroxyalkanolamines used in combination with uracils effect of improving the initial color. This permits production of moldings with service properties complying with expectations over a prolonged period.

16.2.5 Hydrotalcite Clays

Hydrotalcite is a natural mineral. It is the hydroxycarbonate of magnesium and aluminum with the overall formula $Mg_6Al_2(OH)_{16}CO_3 \times 4H_2O$. The structure of hydrotalcite is shown in Figure 16.8.

Hydrotalcite can be produced by a complete synthetic procedure (2). It is intended to replace heavy metal containing stabilizers.

The stabilization activity results from the capacity of the layered double hydroxides to react with the HCl formed during the degradation of PVC. The reaction between the layered double hydroxides and the HCl occurs in a two-step process (3):

1. The counterions between the layers tend to react with the HCl gas, and

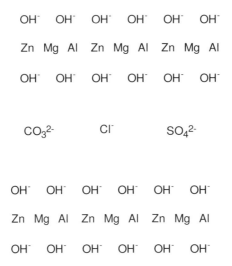

Figure 16.8: Layered Structure of Hydrotalcite

2. The layers themselves react with the HCl under complete destruction of the layered structure and the formation of metal chlorides

The special property when used as nanofiller materials is their thermal decomposition behavior, which makes them also interesting as a potential flame retardant for polymers (4).

16.2.6 Zeolites

Zeolites are effective acid scavengers for halogen containing polymers and enhance the thermal stability of halogen containing polymers. However, the use of zeolites as stabilizers or acid scavengers in halogen containing polymer compounds has been limited for several reasons (5).

First, the zeolites generally have a large particle size, generally in the range of about 3 to about 6 microns. The large size of the zeolite particles not only causes surface blemishes on the finish of the end product made from such a polymer, but also diminishes the physical properties of such polymers. Further, outgassing occurs

Table 16.2: Costabilizers for PVC

Compound Class
Phosphite esters
Epoxy compounds
Poly(ol)s
Phenolic antioxidants
1,3-Diketones
Dihydropyridines
β-Ketocarboxylic acid esters

frequently with polymers containing zeolites when the polymer is heated during processing due to the evolution of water from the zeolite during the heating. As a result, there is foaming.

However, a modified zeolite with a small particle size, a narrow particle size distribution and a water content of less than 10% is useful (5).

16.2.7 Costabilizers

When used alone, costabilizers do not have substantial effect in the thermal stabilization of PVC. However, in combination with stabilizers, mentioned above, they exhibit a synergistic effect. Costabilizers are listed in Table 16.2.

Common epoxy costabilizers are based on epoxidized soya bean oil and tall oil fatty acids or oleic acid.

Poly(ol)s can form complexes and thus may deactivate zinc chloride, etc., which is formed during processing of PVC. This effect leads to a prolongation of the thermal stability of PVC. Poly(ol)s include pentaerythrit, dipentaerythrit, trimethylolpropane, and sorbitol.

Phenolic antioxidants inhibit the autoxidation reactions. This is a general issue of thermal stability, not only restricted to PVC. However, in combination with sulfur containing organotin stabilizers, an increase of effectiveness is observed.

References

1. T. Hopfmann, H.H. Friedrich, K.J. Kuhn, and W. Wehner, Stabilizer system for stabilizing pvc, US Patent 7 358 286, assigned to Chemtura

Vinyl Additives GmbH (Lampertheim, DE), April 15, 2008.
2. F. Cavani, F. Trifiro, and A. Vaccari, Hydrotalcite-type anionic clays: preparation, properties and applications, *Catalysis today*, 11(2):173–301, 1991.
3. L. van der Ven, M.L.M. van Gemert, L.F. Batenburg, J.J. Keern, L.H. Gielgens, T.P.M. Koster, and H.R. Fischer, On the action of hydrotalcite-like clay materials as stabilizers in polyvinylchloride, *Applied Clay Science*, 17(1-2):25–34, July 2000.
4. F.R. Costa, M. Saphiannikova, U. Wagenknecht, and G. Heinrich, Layered double hydroxide based polymer nanocomposites, *Advances in Polymer Science*, 210:101–168, 2008.
5. R.E. Detterman, N.A. Hamerly, C.A. Lepilleur, A.M. Mazany, D.L. Milenius, and A.L. Backman, Halogen containing polymer compounds containing modified zeolite stabilizers, US Patent 6 531 526, assigned to Noveon IP Holdings Corp., March 11, 2003.

17
Acid Scavengers

Acid scavengers are an important type of stabilizers used in the stabilization of poly(vinyl chloride), although they are used in other types of polymers as well. Therefore, this class of substances deserves a separate chapter.

Acid scavengers are also addressed as acid absorbers, antacids, or still less precisely as costabilizers. The addition of acid scavengers is necessary because catalyst residues from processing and manufacture may contribute to undesired properties. This does not necessarily cause a diminished stability of the polymeric base material. Instead, residues from Ziegler-Natta catalysts in poly(olefine)s may contain traces of halogen that could cause corrosion reactions to metals that are in contact with the polymer.

For example, acid scavengers have been proposed as additives in poly(ester)-poly(carbonate) compositions (1).

17.1 Acid Scavenging

Basically, salts of weak acids, such as calcium stearate serve as acid scavengers. They react with strong acids such as hydrogen chloride in the formation of calcium chloride and the generation of the weak acid. The basic principle is shown in Figure 17.1. The acid scavenger is present as an additive in the polymer in amounts of about 25–800 ppm (2).

Calcium lactate compounds may form complexes with trace metals, such as titanium or aluminum. In propylene polymer compositions, potassium citrate as an acid scavenger has been proposed (3). In hydrotalcite type acid scavengers, the chloride anions are inter-

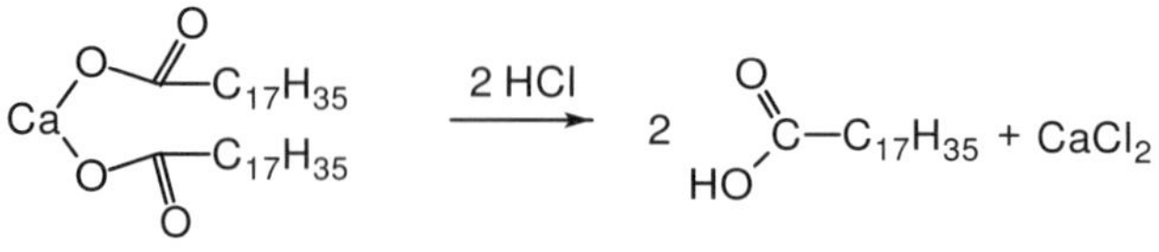

Figure 17.1: Acid Scavenging of Calcium Salts

calated in between the layers. Also zinc oxide is a suitable acid scavengers, as it reacts with hydrogen chloride according to Eq. 17.1.

$$5\,ZnO + 2\,HCl \rightarrow ZnCl_2 + 4\,Zn(OH)_4 \qquad (17.1)$$

17.2 Examples of Formulation

17.2.1 Poly(olefin)s

Most practical uses of propylene polymer compositions require that the composition be stabilized with an acid scavenger. Typically, propylene polymers, which have been formed using catalysts containing halides, require some functionality of an acid scavenger to stabilize the polymer formulation against corrosivity for long-term uses (3).

Zinc stearate and zinc oxide have been proposed as acid scavengers for poly(olefin)s (4). A particularly preferred acid scavenger comprises a mixture of aluminum hydroxide, zinc carbonate and zinc hydroxide (ZHT-4D, Kyowa) (5). Other acid scavengers are hydrotalcites and amorphous basic aluminum magnesium carbonates (6).

Many applications of propylene polymers, such as formation of spun bonded materials and molding applications including thermoforming, preferably require a resin which does not produce large amounts of smoke during processing, and more particularly, does not build up wax like materials around the processing equipment. In many instances, the wax like materials originate from fatty acid derivatives contained in the polymer formulation. Thus, typical acid neutralizers used in conventional poly(propylene) resins, such a calcium stearate which contains long aliphatic chains, e.g., fatty acid salts, should be avoided. One alternative acid neutralizing additive used commercially, calcium lactate, does not produce as much smoke as calcium stearate, but has shown to exhibit screen pack pluggage during melt extrusion of the polymer. This limits production rates of polymer in a commercial facility and may add significant cost to production. Since calcium lactate melts above the temperature of polymer melt extrusion, its particle size is critical in filtering the propylene polymer blended with the additive. A low smoke forming, acid neutralized propylene polymer composition has been proposed that comprises a propylene polymer and mono potassium citrate as acid scavenger (3).

In poly(olefin)s, fluoroelastomers are added sometimes as processing aids. It is believed that the fluoroelastomer and acid scavenger may have an unexpected effect as they modify the blocking and friction properties of the polymer synergistically (4).

17.2.2 Poly(ethylene terephthalate)

In poly(ethylene terephthalate) that is used for beverages, a hydrotalcite-like composition is used as acid scavenger (7). In addition, an acetaldehyde scavenger is used e.g., anthranilamide.

17.2.3 Poly(urethane)s

Examples of useful acid scavengers for poly(urethane)s (PU)s include hydroxides, carbonates, bicarbonates, amines, zeolites, hydrotalcites that will all adequately neutralize acid that may be generated during the PU reaction.

However, some common acid scavengers will adversely affect the PU reaction as they result in materials with poor physical properties. Thus for PUs, more preferred acid scavengers are selected from the group of epoxides and diepoxides (8).

References

1. W.R. Hale, Transparent two phase polyester-polycarbonate compositions, US Patent 7 425 590, assigned to Eastman Chemical Company,

January 18, 2007.
2. A.M. Sukhadia, E.M. Lanier, and L. Moore, Polyethylene compositions, WO Patent 2 007 055 977, assigned to Chevron Phillips Chemical Co. and Sukhadia Ashish M and Lanier Elizabeth M and Moore Louis, May 18, 2007.
3. M.C. Bheda and A. Dorman, Propylene polymer compositions stabilized with potassium citrate as an acid scavenger, US Patent 6 090 877, assigned to BP Amoco Corp., July 18, 2000.
4. A.M. Sukhadia, E.M. Lanier, and L. Moore, Polyethylene compositions, US Patent 7 420 010, assigned to Chevron Philips Chemical Company LP (The Woodlands, TX), September 2, 2008.
5. S.D. Seip, S.E. Thompson, and E.B. Townsend, IV, Polyolefin additive packages for producing articles with enhanced stain resistance, US Patent 6 777 470, assigned to Sunoco, Inc. (R&M) (Philadelphia, PA), August 17, 2004.
6. N. Kaprinidis and N. Lelli, Flame retardant compositions, US Patent 7 109 260, assigned to Ciba Specialty Chemicals Corporation (Tarrytown, NY), September 19, 2006.
7. B. Mahiat and J. Waeler, Acetaldehyde scavenger in polyester articles, WO Patent 2 006 086 365, assigned to Polyone Corp. and Mahiat Bernard and Waeler Jerome, August 17, 2006.
8. S.B. Falloon, R.S. Rose, and M.D. Phillips, Vacuum cooled foams, US Patent 7 008 973, assigned to PABU Services, Inc. (Wilmington, DE), March 7, 2006.

18
Metal Deactivators

In polymers, oxidative degradation is catalyzed strongly by traces of metals that can undergo redox reaction. Most prominent is copper, but also solubilized iron, cobalt, nickel, chromium and manganese may contribute to degradation of the polymers. Apart from polymers, metal deactivators are used widely in lubricating oils.

Many organic materials that are used in electrical technology for insulation purposes, for example, polymers such as (1):

- Polyoxymethylene
- Polyamide
- Unsaturated poly(ester) resins
- Poly(olefin)s

are subject to accelerated thermooxidative aging in the presence of copper. This disadvantageous property considerably impairs the electrical and mechanical properties of these materials in long-term use. The detrimental effect of copper on these materials is particularly aggravated at elevated temperatures since the aging rate of the polymers increases steeply with increasing temperatures.

18.1 Action of Metals in Polymers

Crosslinked polyolefins, which are often used as insulating material for cables and wires, are subject to greatly accelerated aging in the presence of copper and must, be protected effectively against the oxidation accelerating influence of copper (1). The catalytic action of metals is shown in Eq. 18.1.

$$\begin{aligned}
M^{n+} + R-OOH &\rightarrow M^{(n+1)+} + R-O\cdot + HO\cdot + e^- \\
e^- + M^{(n+1)+} + R-OOH &\rightarrow M^{(n)+} + R-OO\cdot + H\cdot \\
HO\cdot + e^- &\rightarrow OH^- \\
H\cdot &\rightarrow H^+ + e^- \\
H^+ + OH^- &\rightarrow H_2O
\end{aligned} \quad (18.1)$$

18.2 Usage

18.2.1 Residues of Catalysts

trans-1,4-Poly(butadiene) is used in tire rubber compounds (2). The green strength of tire rubber compounds can be improved by including *trans*-1,4-poly(butadiene). Moreover, the tire rubber compounds become strain crystallizable. However, due to its high melting point, it is normally necessary to heat *trans*-1,4-poly(butadiene) in order for it to be processed using conventional mixing equipment. This heating step is typically carried out by storing the *trans*-1,4-poly(butadiene) in a at elevated temperatures for a few days prior to its usage. During this storage period, the polymer typically undergoes undesirable oxidative crosslinking which is caused by residual cobalt catalyst and leads to gelation. Actually, the gelation may render the *trans*-1,4-poly(butadiene) unprocessable. For this reason, formulations with metal deactivators have been proposed, in order to avoid this drawback. Salicylic acid, thiosalicylic acid and acetylsalicylic acid, act as metal deactivators in these formulations.

18.2.2 Metallic Reinforcing Parts

Where pneumatic tires, belts, and conveyor belts are in contact with reinforcing metallic elements, tubes are provided with reinforcing cords or wires.

In general, in the production of all rubber articles in which rubber is reinforced with metal, it is necessary to obtain between the metal and the elastomeric composition a strong and durable bond in order

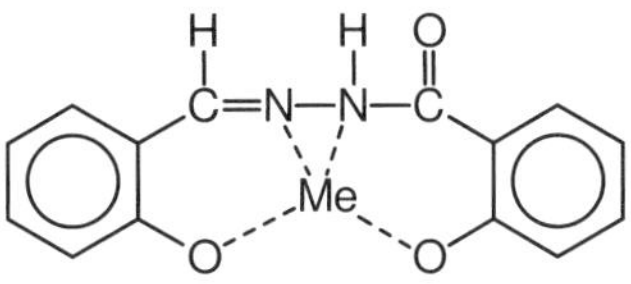

Figure 18.1: Metal Complex

to ensure a good efficiency and a long life for the articles produced (3).

One solution to this problem of oxidative degradation has been to add compounds which deactivate the metals, particularly copper, thereby suppressing their catalytic capacity and substantially reducing oxidative degradation. Metal deactivators act by complexing the respective metals as shown in Figure 18.1.

18.3 Examples of Metal Deactivators

Examples of metal deactivators are shown in Figure 18.2. Selected metal deactivators are shown in Table 18.1.

N,N'-bis(salicyloyl)hydrazide can be prepared through the reaction of a salicylic acid alkyl ester and hydrazine. Alternatively, a thionyl chloride route is viable (4). Detailed examples of the synthesis are given in the literature (5).

N,N'-Bis(salicyloyl)hydrazide serves particularly as a metal deactivator in cable and wire insulation for power and communications engineering (1). This compound allows a non-hazardous processing in engineering applications due to a substantially reduced eye irritation in comparison to other metal deactivators.

In addition to low molecular compounds, polyhydrazides of 3,3'-thiodipropionylhydrazide have been proposed as metal deactivators (6). These compounds can be manufactured by the reaction of a 3,3'-thiodipropionic acid dihydrazide with a suitable bis(acid chloride).

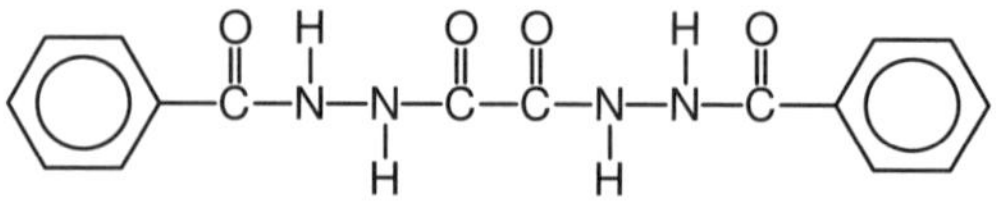

N,*N*′-Dibenzoyloxalic acid dihydrazide

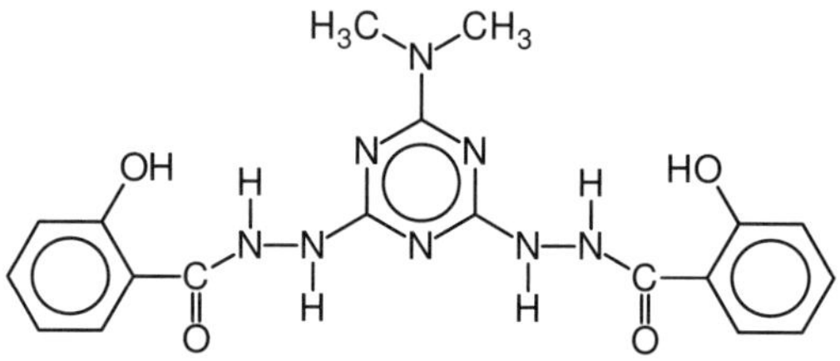

2-(Dimethylamino)-4,6-disalicyloylhydrazine-triazine

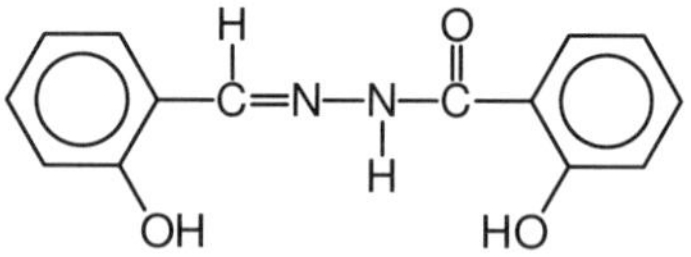

N-salicylidene-*N*′-salicylhydrazide

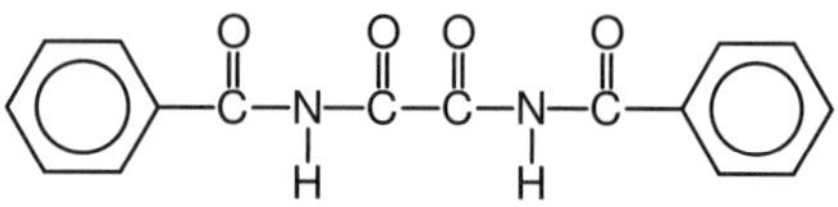

N,*N*′-Diphenyloxamide

Figure 18.2: Metal Deactivators (3)

Table 18.1: Metal Deactivators (7)

Compound
N,N'-Diphenyloxamide
N,N-dibenzoylhydrazine (5)
N-Salicylal-N'(salicyloyl)hydrazide
N,N'-Bis(salicyloyl)hydrazide (5)
N,N'-Bis(3,5-di-*tert*-butyl-4-hydroxyphenylpropionyl)hydrazine
3-N-Salicyloylamino-1,2,4-triazole (Adekastab® CDA-1) (8)
Decamethylene dicarboxylic acid-bis(N'-salicyloylhydrazide) (Adekastab® CDA-6) (8)
Bis(benzylidene)oxalyl dihydrazide
Oxanilide
Isophthaloyl dihydrazide
Sebacoyl bisphenylhydrazide
N,N'-diacetyladipoyl dihydrazide
N,N'-bis(salicyloyl)oxalyl dihydrazide
N,N'-bis(salicyloyl)thiopropynyl dihydrazide
2',3-Bis[[3-[3,5-di-*tert*-butyl-4-hydroxyphenyl]propionyl]]propionohydrazide (Irganox® MD 1024) (9)
1,2-Bis(3,5-di-*tert*-butyl-4-hydroxyhydrocinnamoyl)hydrazine (10)

18.3.1 Side Effects

Metal deactivators may have side effects. For example, they act additionally as adhesion promoters in the vulcanization of elastomers (3, 11).

Linear low density poly(ethylene)-based nanocomposites exhibit a faster photo oxidation in comparison to the unfilled matrix. The acceleration is not due to a faster rate of the photo oxidation but due to the reduction of the induction time of the oxidation reaction. It is suspected that the presence of trace amounts of metal ions in the organoclays promotes a catalytic photo oxidation. Thus, metal deactivators have been introduced into the formulations. Combinations of metal deactivators with UV absorbers show synergistic effects (12).

For styrene-butadiene copolymers, extended experiments with different formulations, including a metal deactivator have been published (13). A metal deactivator, such as Irganox® MD 1024, in combination of Irgafos® 168 or Irganox® 1010 was found to be antagonistic. The antagonistic effect between phenolic antioxidants and Irganox® MD 1024 may arise in that the metal deactivator bears a hindered phenolic group and behaves in the same way as a phenolic antioxidant.

References

1. W. von Gentzkow and R. Rubner, N,N'-Bis-salicyloyl-hydrazine as a metal deactivator, US Patent 4 446 266, assigned to Siemens Aktiengesellschaft (Munich, DE), May 1, 1984.
2. K.F. Castner, Metal deactivator for cobalt catalyzed polymers, US Patent 6 013 736, assigned to The Goodyear Tire & Rubber Company (Akron, OH), January 11, 2000.
3. G.K. Cowell and D.J. Cherry, Metal deactivators as adhesion promotors for vulcanizable elastomers to metals, US Patent 3 994 987, assigned to Ciba-Geigy Corporation (Ardsley, NY), November 30, 1976.
4. P.E.R.E.J. Lakeman, T.H. Ho, and R. Wevers, Heavy metal deactivator/inhibitor for use in olefinic polymers, WO Patent 2 008 127 830, assigned to Dow Global Technologies Inc., Lakeman Pascal E R E J, Ho Thoi H, and Wevers Ronald, October 23, 2008.
5. T. Yoshikawa, N. Sakamoto, M. Kurita, S. Oh-e, and L. Nagamori, Stabilized olefin polymer composition, US Patent 4 043 976, assigned to UBE Industries, Ltd. (Ube, JA), August 23, 1977.
6. R.H.S. Wang and G. Irick, Jr., Polyhydrazides as stabilizers for polyolefins, US Patent 4 087 405, assigned to Eastman Kodak Company (Rochester, NY), May 2, 1978.
7. P. Piccinelli, M. Vitali, A. Landuzzi, G. Da Roit, P. Carrozza, M. Grob, and N. Lelli, Permanent surface modifiers, US Patent 7 408 077, assigned to Ciba Specialty Chemicals Corp. (Tarrytown, NY), August 5, 2008.
8. Y. Oobayashi and K. Kitano, Fiber-reinforced polyolefin resin composite and molded article obtained from the same, US Patent 6 780 506, assigned to Sumitomo Chemical Company, Limited (Osaka, JP), August 24, 2004.
9. S. Morlat-Therias, E. Fanton, J.L. Gardette, N.T. Dintcheva, F.P. La Mantia, and V. Malatesta, Photochemical stabilization of linear low-density polyethylene/clay nanocomposites: Towards durable nanocomposites, *Polymer Degradation and Stability*, 93(10):1776–1780, October 2008.
10. E.H. Jancis, Thermoplastic resins in contact with metals or metal salts stabilized by blends of dithiocarbamates and metal deactivators, US Patent 6 790 888, assigned to Crompton Corporation (Middlebury, CT), September 14, 2004.

11. G.K. Cowell and D.J. Cherry, Metal deactivators as adhesion promotors for vulcanizable elastomers to metals, US Patent 4 306 930, assigned to Ciba-Geigy Corporation (Ardsley, NY), December 22, 1981.
12. F. La Mantia, N.T. Dintcheva, V. Malatesta, and F. Pagani, Improvement of photo-stability of LLDPE-based nanocomposites, *Polymer Degradation and Stability*, 91(12):3208–3213, December 2006.
13. N.S. Allen, A. Barcelona, M. Edge, A. Wilkinson, C. Galan Merchan, and V. Ruiz Santa Quiteria, Aspects of the thermal and photostabilisation of high styrene-butadiene copolymer (SBC), *Polymer Degradation and Stability*, 91(6):1395–1416, June 2006.

19
Oxidative Degradation

19.1 Autoxidation

Autoxidation was first investigated in the aging of natural rubber, although it plays a rather general role. Autoxidation starts with the formation of free radicals delivered from the backbone of the polymer.

$$\begin{aligned} R\text{--}H &\rightarrow R\cdot \\ R\text{--}R &\rightarrow R\cdot \end{aligned} \quad (19.1)$$

Subsequently, in presence of oxygen a chain propagation is initiated.

$$\begin{aligned} R\cdot + O_2 &\rightarrow R\text{--}O\text{--}O\cdot \\ R\text{--}O\text{--}O\cdot + R\text{--}H &\rightarrow R\text{--}O\text{--}O\text{--}H + R\cdot \end{aligned} \quad (19.2)$$

Besides the chain propagation, chain branching reactions may occur. Branching reactions increase the number of reactive radicals. This type of reactions plays a role in explosion reactions. In weaker cases, branching effects an apparently effective acceleration of the reaction under consideration.

$$R\text{--}O\text{--}O\text{--}H \rightarrow R\text{--}O\cdot + \cdot O\text{--}H \quad (19.3)$$

The reaction is stopped by termination sequences. Labile peroxides may be formed that can decompose further.

$$\begin{aligned} R\cdot + R\text{--}O\text{--}O\cdot &\rightarrow R\text{--}O\text{--}O\text{--}R \\ R\cdot + R\cdot &\rightarrow R\text{--}R \end{aligned} \quad (19.4)$$

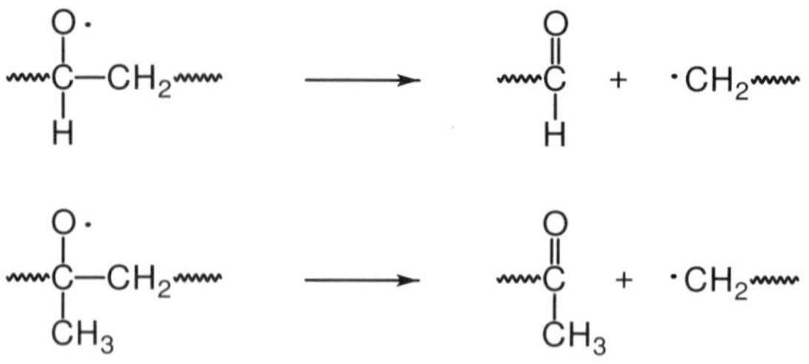

Figure 19.1: β-Scission

Figure 19.2: Formation of Alkyl Radicals

β-Scission plays an important role in the mechanism of degradation. This mechanism is shown in Figure 19.1.

Alkyl radicals may be formed by a mechanism depicted in Figure 19.2. The formation of alkyl radicals occurs predominately, when shear is applied at elevated temperatures, i.e., during processing.

19.2 Inhibition of Autoxidation

The inhibition of autoxidation may be achieved by the addition of:

- Radical scavengers
- Hydrogen donors
- Redox catalysts
- Metal deactivators.

Trapping of radicals by scavengers belongs to the most important methods to prevent autoxidation reactions. Radical scavengers are also referred to as chain breaking acceptors. Further, they are classified as primary antioxidants.

Figure 19.3: Decomposition of Peroxide Radicals by Aromatic Amines

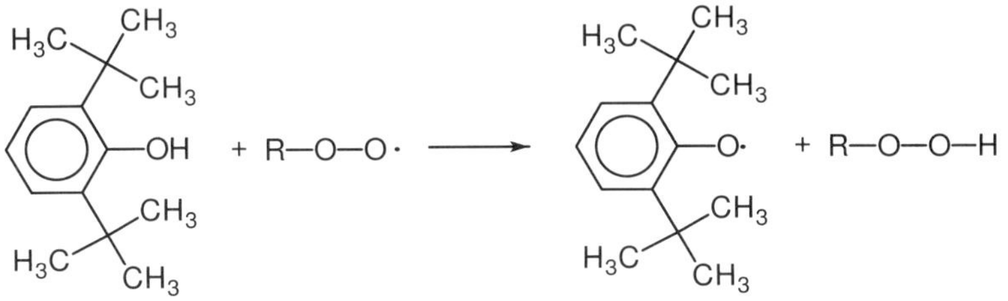

Figure 19.4: Decomposition of Peroxide Radicals by Sterically Hindered Phenols

Hydrogen donors are referred to as chain breaking donors. In particular, hydrogen donors decompose peroxides into inert products. Hydrogen donors are classified as primary antioxidants, because they access secondary products in the chain of the autoxidation reaction. The mechanism that interferes the autoxidation reaction is shown in Eq. 19.5.

$$\begin{aligned} ROO\cdot + R{-}H &\rightarrow ROOH + R\cdot \\ ROO\cdot + In{-}H &\rightarrow ROOH + In\cdot \\ In\cdot + R{-}H &\rightarrow In{-}H + R\cdot \end{aligned} \quad (19.5)$$

Special types of hydrogen donors include phenols and aromatic amines. Notably, aromatic amines are also known as accelerators in peroxide curing of unsaturated poly(ester)s resins. The mechanism of the decomposition of peroxide radicals by aromatic amines is shown in Figure 19.3.

In the same manner, sterically hindered phenols interact with peroxide radicals as they are forming stable radicals. The mechanism of interaction is shown in Figure 19.4.

Hydroperoxides are still reactive in another manner as they may form peroxide radicals. Hydroperoxide decomposers react with

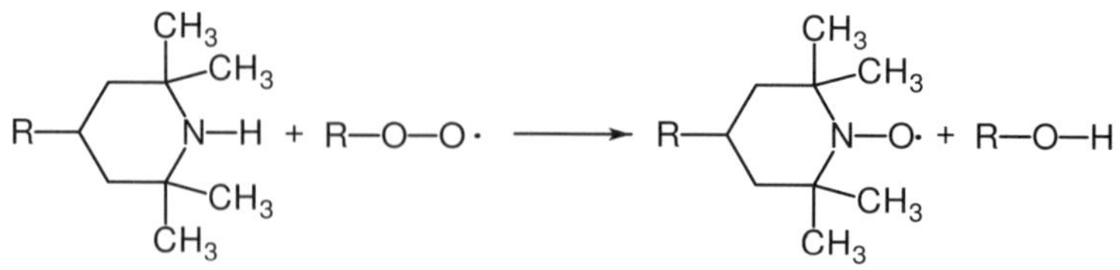

Figure 19.5: Formation of Stable Nitroxyl Radicals

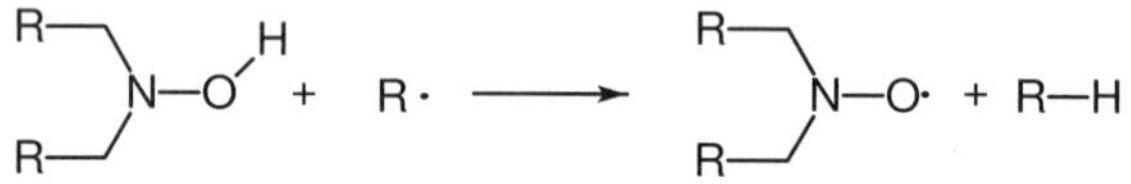

Figure 19.6: Formation of Stable Nitroxyl Radicals

hydroperoxides as they transform them into stable products. The reaction is preferably a redox reaction. For example, phosphites or phosphonites are oxidized into phosphates as shown in Eq. 19.6.

$$\begin{aligned} H_2P(-O\Phi)_3 + R-OOH &\rightarrow O=P(-O\Phi)_3 + R-OH \\ H_2P(-O\Phi)_3 + R-OO\cdot &\rightarrow O=P(-O\Phi)_3 + R-O\cdot \\ H_2P(-O\Phi)_3 + R-O\cdot &\rightarrow O=P(-O\Phi)_3 + R\cdot \end{aligned} \quad (19.6)$$

Thio compounds are active in the same way. Eq. 19.7 shows the mechanism of peroxide decomposition using thio compounds.

$$H-S(-R)_2 + R-OOH \rightarrow O=S(-H)(-R)_2 + R-OH \quad (19.7)$$

Hindered amine stabilizers act as scavengers for alkyl radicals. They act as they are forming stable nitroxyl radicals. The mechanisms of action is shown in Figure 19.5.

In addition, hydroxyl amines are capable of forming stable nitroxyl radicals. The mechanisms of action is shown in Figure 19.6. Actually, the reactive species is an intermediately formed nitrone, $>N^+-O^-$.

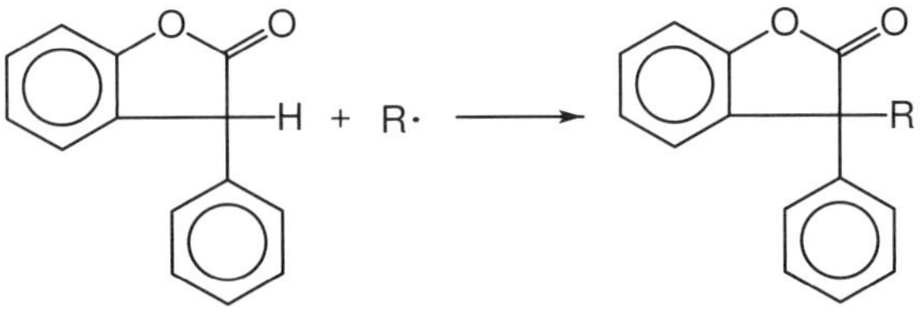

Oxidative Degradation

Figure 19.7: Benzofuran Derivatives as Radical Scavengers

Benzofuran derivatives are highly active radical scavengers. They are used as processing aids. In particular the hydrogen close to the carboxyl group is active in scavenging. The mechanism is shown in Figure 19.7.

Phosphite and phosphonite esters act as antioxidants by three basic mechanisms depending on their structure (1). Basically, phosphites and phosphonites are secondary antioxidants that decompose hydroperoxides. Their performance in hydroperoxide decomposition decreases from phosphonites, alkyl phosphites, aryl phosphites, down to hindered aryl phosphites. Five membered cyclic phosphites act catalytically by the formation of acidic hydrogen phosphates. In contrast, hindered aryl phosphites are interrupting the branched kinetic chain.

The change in the state of oxidation occurs typically in a redox reaction. Often, redox reactions are catalyzing the decomposition of peroxides. Certain metals may occur in varying states of oxidation. These metals may be readily involved in redox reactions that eventually decompose peroxides. In practice, most notorious is copper and further iron. The mechanism or the redox catalyzed reaction is shown in Eq. 19.8.

$$M^{n+} + R\text{-}OOH \rightarrow M^{(n+1)+} + R\text{-}O\cdot + HO\cdot + e^-$$
$$e^- + M^{(n+1)+} + R\text{-}OOH \rightarrow M^{(n)+} + R\text{-}OO\cdot + H\cdot$$
(19.8)

Thus, as the metal is regenerated in its original state of oxidation, the mechanism is a pure catalytic mechanism.

Metal deactivators act as complex forming agents. When the metal is complexed, it loses its ability to undergo a redox reaction. An example of such a complex is shown in Figure 19.8.

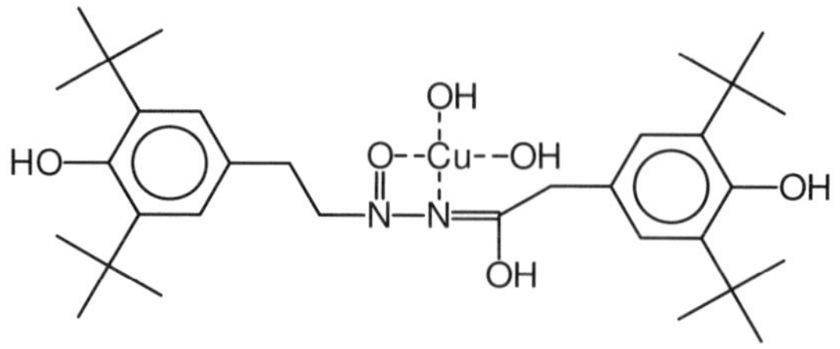

Figure 19.8: Metal Complex Formation

Table 19.1: Classes of Antioxidants (2)

Compound class
Alkylated Monophenols
Alkylthiomethylphenols
Hydroquinones
Tocopherols
Hydroxylated thiodiphenyl ethers
Alkylidenebisphenols
O-, N- and S-Benzyl compounds
Hydroxybenzylated malonates
Aromatic hydroxybenzyl compounds
Triazine compounds
Benzyl phosphonates
Acylaminophenols
β-(5-*tert*-Butyl-4-hydroxy-3-methylphenyl)propionic acid esters
Aminic antioxidants

There are a lot of antioxidants on the market. Several classes of antioxidants have been described, which care summarized in Table 19.1. A more detailed listing is given in the literature (3,4) and in Table 19.2.

Some steric hindered phenol antioxidants are shown in Figure 19.9. Thiophenols are shown in Figure 19.10. Tocopherol is shown in Figure 19.11. Triazine based antioxidants are shown in Figure 19.12.

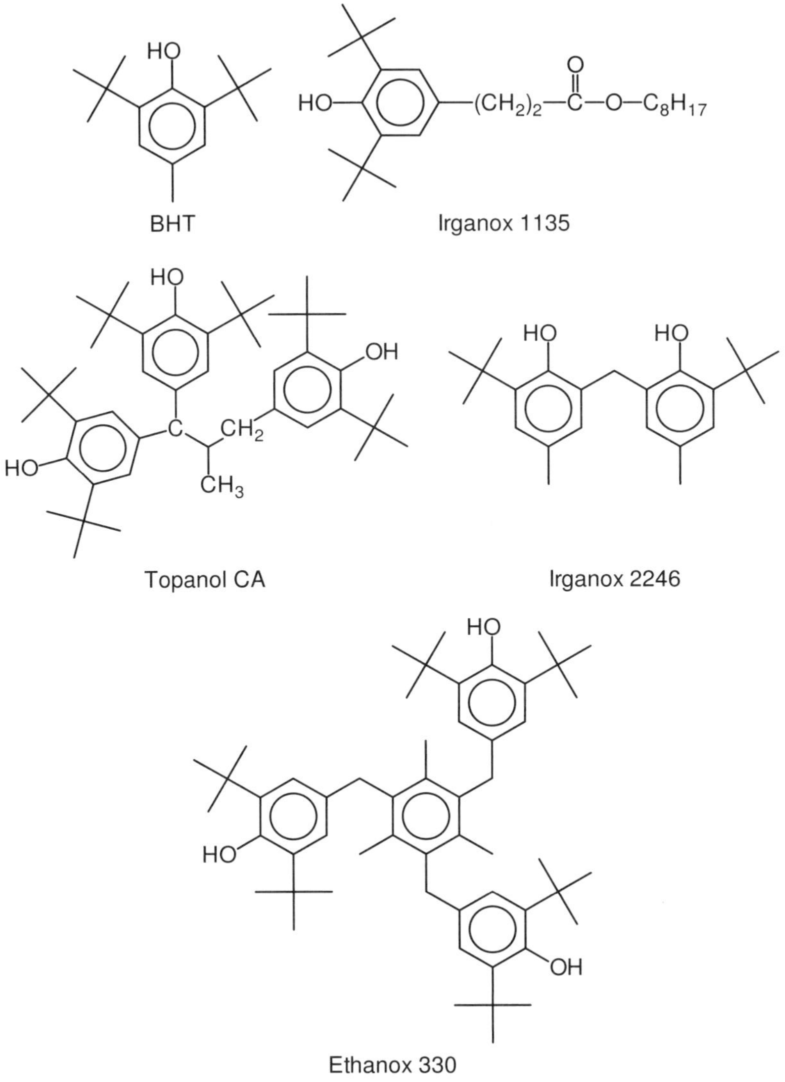

Figure 19.9: Steric Hindered Phenol Antioxidants

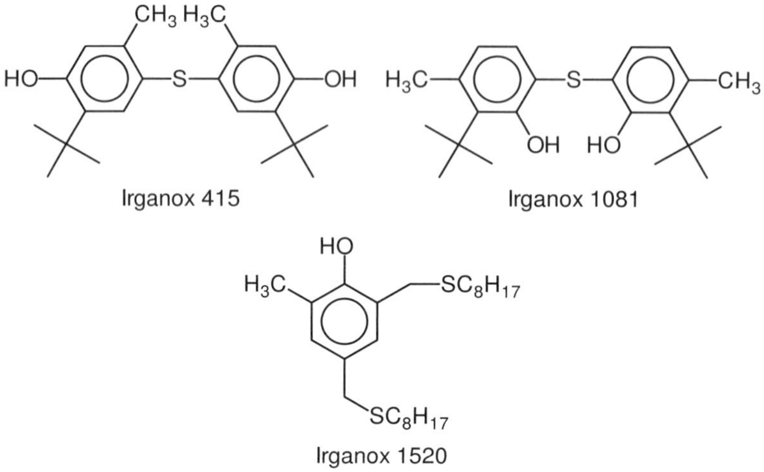

Figure 19.10: Thiophenol Antioxidants

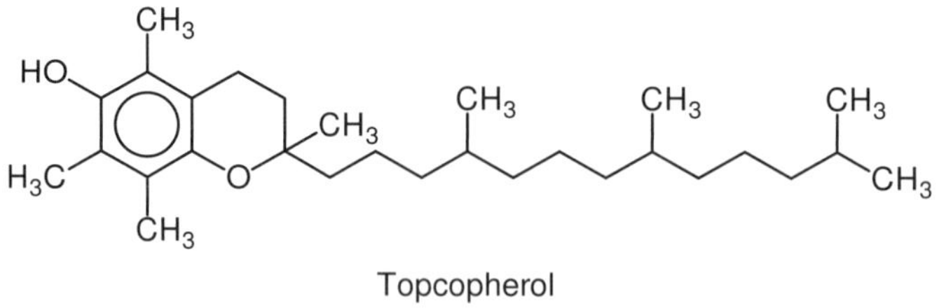

Figure 19.11: Tocopherol Based Antioxidants

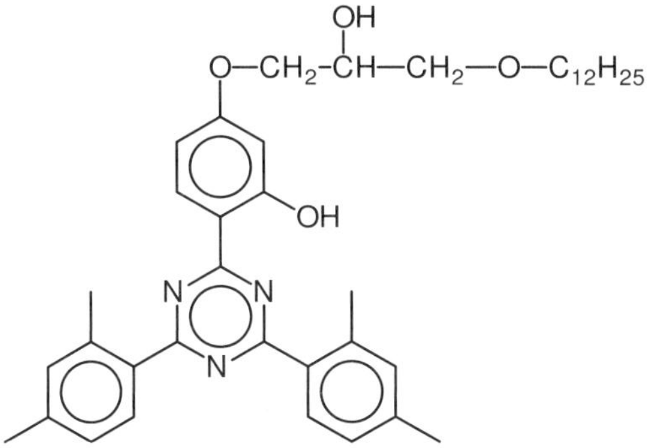

Tinuvin 400

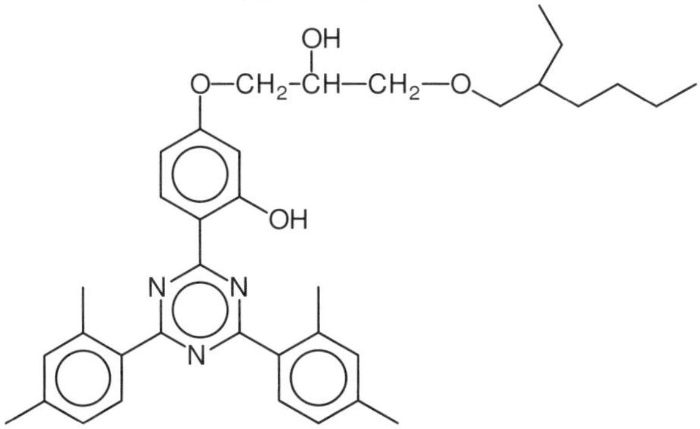

Tinuvin 405

Figure 19.12: Triazine Based Antioxidants

Table 19.2: Antioxidants (2)

Compounds
Alkylated Monophenols
2,6-Di-*tert*-butyl-4-methylphenol (BHT) 1,1,3-Tris(2-methyl-4-hydroxy-5-*tert*-butylphenyl)butane (Topanol CA) Irganox® 1135 Irganox® 2246 Ethanox® 330 2,6-Di-*tert*-butyl-*p*-cresol (5) Stearyl(3,3-dimethyl-4-hydroxybenzyl) thioglycolate (5)
Alkylthiomethylphenols
Irganox® 415 Irganox® 1081 2,4-Dioctylthiomethyl-6-methylphenol (Irganox® 1520)
Tocopherols
α-Tocopherol
Triazine compounds
Tinuvin® 400 Tinuvin® 405
Cinnamates
Tetrakis[methylene (3,5-di-tert-butyl-4-hydroxy)hydrocinnamate]
Phosphites
Triphenyl phosphite Tris(nonylphenyl) phosphite Trilauryl phosphite

References

1. K. Schwetlick and W.D. Habicher, Organophosphorus antioxidants action mechanisms and new trends, *Angewandte Makromolekulare Chemie*, 232(1):1522–9505, January 1995.
2. P. Piccinelli, M. Vitali, A. Landuzzi, G. Da Roit, P. Carrozza, M. Grob, and N. Lelli, Permanent surface modifiers, US Patent 7 408 077, assigned to Ciba Specialty Chemicals Corp. (Tarrytown, NY), August 5, 2008.
3. H. Zweifel, ed., *Plastics Additives Handbook*, Hanser Publishers, Munich, 5th edition, 2001.
4. H. Zweifel, R.D. Maier, and M. Schiller, eds., *Plastics Additives Handbook*, Hanser Publishers, Munich, 6th edition, 2009.
5. T. Tanizaki, T. Nakahara, and M. Kubo, Poly (4-methyl-1-pentene) resin laminates and uses thereof, US Patent 6 265 083, assigned to Mitsui Chemicals, Inc. (Tokyo, JP), July 24, 2001.

20
Degradation by Light

The degradation of polymers by light is not only a photodegradation, but rather a photooxidative degradation, since in practice the degradation occurs in the presence of air. Thus, secondary degradation reactions will occur where oxygen is involved.

An important exception of this view in the photodegradation under the conditions where space crafts are operating. In contrast, the type of radiation is more severe still, since under normal atmospheric conditions a lot of hard radiation is filtered by the atmosphere.

When light enters a surface of a material, it may by either reflected from the surface, scattered, pass unchanged, or absorbed in the interior of the material.

Photochemistry teaches that absorbed light causes photochemical reactions, such as crosslinking or degradation.

In particular, the absorption of light is related to the chemical structure of the material. Groups that interact strongly with light are addressed as chromophores. Well known chromophores include carbon-carbon double bonds, and double bonds where hetero atoms are involved. Double bonds are still more sensitive to degradation. Carbonyl double bonds absorb in the spectral region above 290 nm. Besides the absorption spectrum, the sensitivity to radiation depends on the bond strength. Bonds that can be easily broken can be expected to undergo a more pronounced degradation than bonds with a high energy content.

Chromophores may not necessarily be attached to the basic structure of the polymer under consideration. Chromophores may be introduced as impurities in the course of manufacture. Examples are catalyst residues and other processing aids.

The sensitivity of chromophores is dependent on the wavelength of the incident light. For this reason, some spectral parts of the sunlight are increasingly effective of photodegradation while other parts are less effective. Light of varying wavelength may even cause a different reaction pattern.

This pattern may be vary with the type of material. For example, when poly(ethylene) (PE) and poly(propylene) (PP) are compared to incident light of a normalized intensity, at wavelengths below 330 nm, PP is more sensitive in comparison to PE, whereas wavelengths above 330 nm the situation is reverse. In testing, it has been found that xenon arcs and natural sunlight show very similar results. Therefore, weathering tests using artificial light can be designed.

The stability of various polymers against UV and weathering have been described extensively in the literature (1). In semi-crystalline materials light is scattered in multiple paths. Thus, the average path through the material is increased considerably. This phenomenon leads to an increased sensitivity of crystalline polymers to light.

The degradation reaction starts with the electronic excitation of the bonding electrons. The very types of excitation can be visualized in the Jablonsky diagram, which is shown in Figure 20.1.

When light is absorbed, the electronic state changes from the ground state into an excited single state. Thereafter, several processes may occur.

Most simply, the light absorbed resulting to the singlet state is released. Alternatively, after a series of vibrational relaxation processes, the energy is released with a different wavelength from the singlet state. This process is referred to as fluorescence. Likewise, after an internal conversion the energy may be released by a series of vibrational relaxation processes.

In another type of process, if a triplet state is available at the same energy as the singlet state, an intersystem crossing process may occur into the triplet state. From there, as in the singlet state, a series of vibrational relaxation processes can occur and the light is released again with a different wavelength from the triplet state. This process is referred to as phosphorescence.

In comparison to fluorescence, has the process of phosphorescence more individual steps involved. This explains qualitatively why the fluorescence phenomenon is much faster than phosphores-

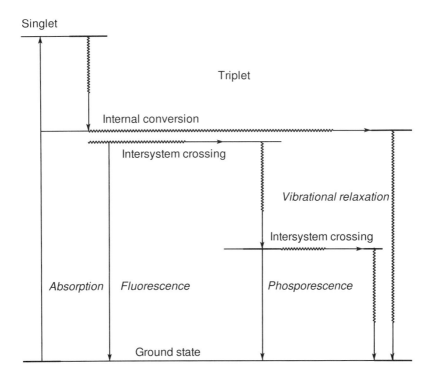

Figure 20.1: Jablonsky Diagram

Table 20.1: Light and Bond Energy (2, p.151)

Wavelength [nm]	Energy [kJ Einstein^{-1}]	Bond	Energy [kJ mol^{-1}]
290	419	C–H	380–420
300	400	C–C	340–350
320	375	C–O	320–380
350	340	C–Cl	300–340
400	300	C–N	320–330

cence. From the triplet state, analogously to the singlet state, an intersystem crossing may occur, followed by a series of vibrational relaxation processes reaching eventually the ground state.

In Table 20.1 bond energies of characteristic bonds are shown, as well as the corresponding energies of light with the particular wavelength.

The unit Einstein refers to one mol of photons. The bond energy is the energy to separate the bond into two separate atoms bearing radicals, i.e., single unpaired electrons. In other words, we are talking about the homolytic cleavage of the bond. The energy E is related to the frequency ν via the well known and famous relation

$$E = h\nu. \tag{20.1}$$

Here h is Planck's constant. The wavelength λ is related to the frequency via the velocity of light c as

$$\lambda\nu = c. \tag{20.2}$$

The absorption of light follows the law of Lambert and Beer. The transmitted intensity of light I with respect to the intensity of the entering light I_0 is

$$\ln\frac{I}{I_0} = -\epsilon l c. \tag{20.3}$$

Here, l is the length of the path, c is the concentration of the absorbing species, and ϵ is the coefficient of extinction with respect to the particular absorbing species.

20.1 Photolysis

In contrast to autoxidation, the initial reactions differ in photolysis. Reactive oxygen radicals are formed by the photolysis of ozone or nitrogen dioxide. Some air pollutants such as O_3 and NO_2 enhance the photolysis.

$$\begin{aligned} O_3 &\rightarrow O_2 + \cdot O\cdot \quad (h\nu < 590\,\text{nm}) \\ NO_2 &\rightarrow NO + \cdot O\cdot \quad (h\nu < 420\,\text{nm}) \end{aligned} \quad (20.4)$$

In particular, NO_2 undergoes a catalytic cycle, as sketched out in Eq. 20.5.

$$\begin{aligned} NO_2 &\rightarrow NO + \cdot O\cdot \\ NO + R{-}OO\cdot &\rightarrow NO_2 + R{-}O\cdot \end{aligned} \quad (20.5)$$

Thereafter, the atomic oxygen reacts with hydrogens from the polymer backbone.

$$\begin{aligned} \cdot O\cdot + R{-}H &\rightarrow \cdot O{-}H + R\cdot \\ \cdot O{-}H + R{-}H &\rightarrow H_2O + R\cdot \\ O_2 + R\cdot &\rightarrow R{-}OO\cdot \end{aligned} \quad (20.6)$$

Important photochemical reactions are the Norrish Type I and the Norrish Type II reaction. These reactions are depicted in Figure 20.2.

Recent studies have shown that tensile and shear stresses accelerate the rate of the photochemical degradation of polymers. In contrast, compressive stress generally retards the rate of photochemical degradation (3, 4). In addition to stress, other factors may affect the rates of polymer photodegradation. These factors include the:

- Intensity of absorbed light
- Morphology of the polymer
- Rate of oxygen diffusion in the polymer
- Concentration of the chromophores.

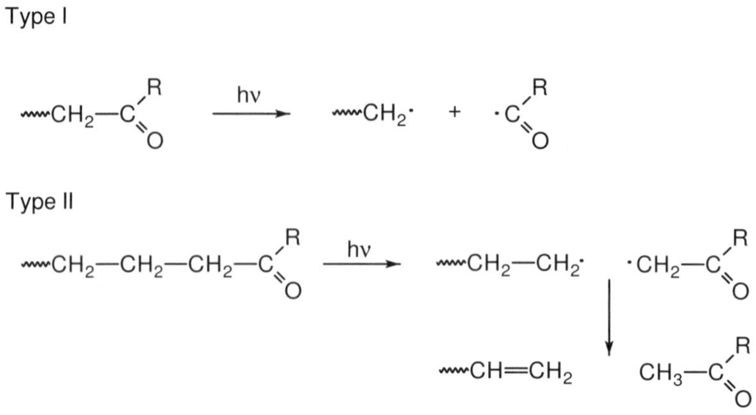

Type I

$$\text{\textasciitilde CH}_2-\underset{\underset{O}{\parallel}}{C}-R \xrightarrow{h\nu} \text{\textasciitilde CH}_2\cdot \; + \; \cdot\underset{\underset{O}{\parallel}}{C}-R$$

Type II

$$\text{\textasciitilde CH}_2-\text{CH}_2-\text{CH}_2-\underset{\underset{O}{\parallel}}{C}-R \xrightarrow{h\nu} \text{\textasciitilde CH}_2-\text{CH}_2\cdot \quad \cdot\text{CH}_2-\underset{\underset{O}{\parallel}}{C}-R$$

$$\downarrow$$

$$\text{\textasciitilde CH}=\text{CH}_2 \qquad \text{CH}_3-\underset{\underset{O}{\parallel}}{C}-R$$

Figure 20.2: Norrish Reactions

20.1.1 Redox Catalysis

Metals can accelerate the decomposition of hydroperoxides that are formed e.g., by Eq. 19.2.

$$\begin{aligned} M^{n+} + R-O-O-H &\rightarrow M^{(n+1)+} + R-O\cdot + OH^- \\ M^{(n+1)+} + R-O-O-H &\rightarrow M^{n+} + R-O-O\cdot + H^+ \\ H^+ + OH^- &\rightarrow H_2O \end{aligned} \qquad (20.7)$$

The second reaction in Eq. 20.7 regenerates the original metal ion M^{n+}. In the combined reaction, the hydroxyl anion and the hydrogen cation recombine to water. This reaction is used purposely in hydroperoxide curing of unsaturated poly(ester) resins.

20.1.2 Scavenging

One important mechanism in order to delay degradation reactions is the addition of scavengers bearing S–H moieties. These groups react with the highly reactive radicals as they convert them into quasi inert radicals S·.

$$S-H + \cdot O-H \rightarrow S\cdot + H_2O \qquad (20.8)$$

Cyclic mechanisms have been described with stable radicals in hindered amine stabilizers.

$$\begin{aligned} >\!N\!-\!O\cdot + R\cdot &\rightarrow\; >\!N\!-\!O\!-\!R \\ >\!N\!-\!O\!-\!R + R\!-\!O\!-\!O\cdot &\rightarrow\; >\!N\!-\!O\cdot + R\!-\!O\!-\!O\!-\!R \end{aligned} \quad (20.9)$$

For example, in Eq. 20.9, the $>\!N\!-\!O\cdot$ unit is regenerated, as simultaneously peroxides are formed.

20.2 Photooxdation

We consider now secondary reactions of oxidation. These reactions are typically taking place in PE. First of all, some radicals are created by the process of photolysis. Let them be

$$\begin{aligned} &\rightarrow\; R\cdot \\ &\rightarrow\; RO\cdot \\ &\rightarrow\; HO\cdot \\ &\rightarrow\; HOO\cdot \end{aligned} \quad (20.10)$$

As secondary reactions we have a chain propagation, like

$$\begin{aligned} R\cdot + O_2 &\rightarrow\; R\!-\!OO\cdot \\ R\!-\!OO\cdot + R\!-\!H &\rightarrow\; R\!-\!OOH + R\cdot \end{aligned} \quad (20.11)$$

The rate of reaction is considerably accelerated by branching reactions,

$$\begin{aligned} R\!-\!OOH &\rightarrow\; R\!-\!O\cdot + \cdot OH \\ R\!-\!OOH + R\!-\!H &\rightarrow\; R\!-\!O\cdot + R\cdot + H_2O \\ R\!-\!OOH + R\!-\!OOH &\rightarrow\; R\!-\!OO\cdot + R\!-\!O\cdot + H_2O \end{aligned} \quad (20.12)$$

Eventually, the chain is terminated by reactions like

$$\begin{aligned} R\cdot + R\cdot &\rightarrow\; R\!-\!R \\ R\!-\!OO\cdot + R\cdot &\rightarrow\; R\!-\!OO\!-\!R \\ R\!-\!OO\cdot + R\!-\!OO\cdot &\rightarrow\; >\!R\!=\!O + R\!-\!OH \\ R\!-\!OO\cdot + R\!-\!OO\cdot &\rightarrow\; R\!-\!OO\!-\!R + O_2 \end{aligned} \quad (20.13)$$

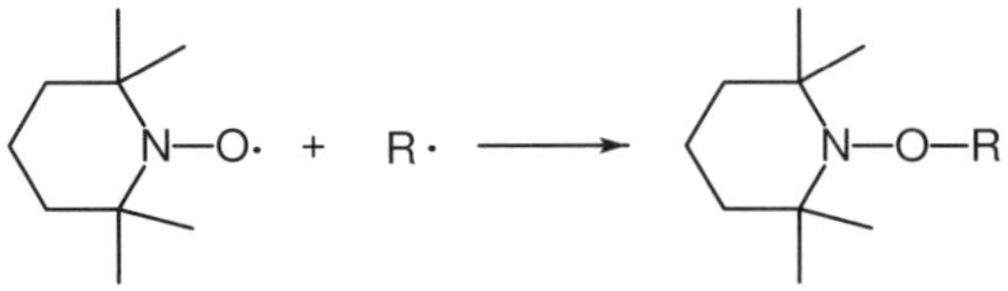

Figure 20.3: Formation of Stable Nitroxyl Radicals

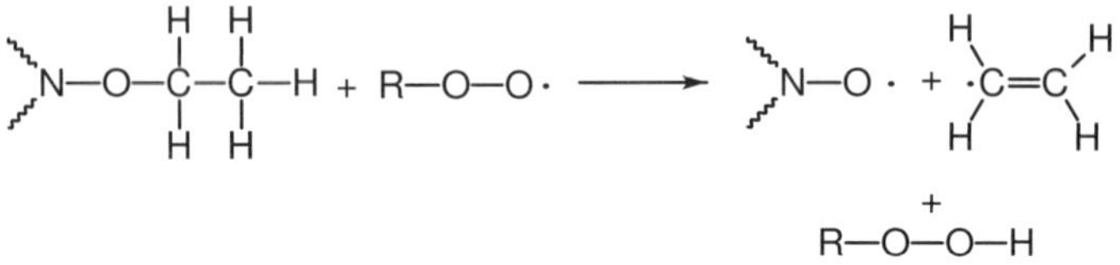

Figure 20.4: Formation of Stable Nitroxyl Radicals

Hindered amine light stabilizers are reacting via the nitroxyl function. The mechanism of action is shown in Figure 20.3. Examples of reactions of nitroxyl radicals are shown in Eq. 20.14.

$$\begin{aligned}
>\text{N-H} + \text{O}_3 &\rightarrow\; >\text{N-O}\cdot \\
>\text{N-H} + \text{H}_2\text{O}_2 &\rightarrow\; >\text{N-O}\cdot \\
>\text{N-H} + \text{R-OO-H} &\rightarrow\; >\text{N-O}\cdot \\
>\text{N-O}\cdot + \text{R}\cdot &\rightarrow\; >\text{NOH} \\
>\text{N-O}\cdot + \text{R-H} &\rightarrow\; >\text{NOH} + \text{R}\cdot
\end{aligned} \qquad (20.14)$$

The basic structure of hindered amine light stabilizers consists of 2,2,6,6-tetramethylpiperidine. The term hindered refers to the four methyl groups shielding the nitrogen atom. If the structure is hindered less, considerable loss in efficiency is observed.

Another related class to sterically hindered amines are sterically hindered amine ethers. These compounds are equally capable of forming nitroxyl radicals, as shown in Figure 20.4. Moreover, for example, the ethyl ethers decompose thermally into a hydroxylamine compound and ethene.

Table 20.2: Classes of UV Stabilizers

Stabilizer	Remark
2-Hydroxybenzophenones	
2-hydroxyphenylbenzotriazoles	
Sterically hindered amines	
Salicylates	
Cinnamate derivatives	
Resorcinol monobenzoates	
Oxanilides	
p-hydroxybenzoates	

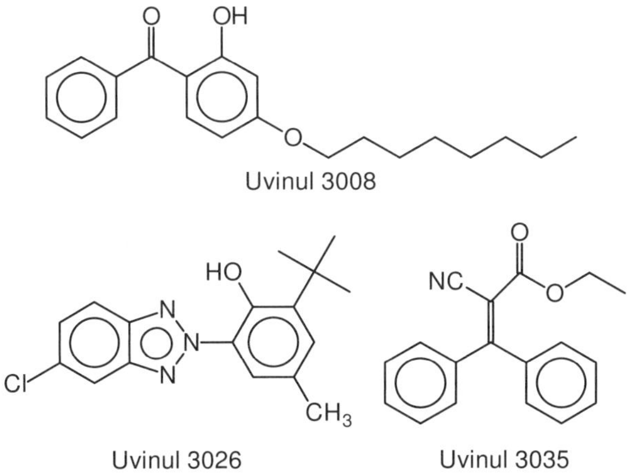
O
OH
O
Uvinul 3008
HO
N
N
N
Cl
CH3
Uvinul 3026
O
NC
O
Uvinul 3035

Figure 20.5: Light Stabilizers (5)

20.3 UV Stabilizers

The most important classes of UV stabilizers are summarized in Table 20.2. Details are given in Table 20.3. Some selected commercial available products are described in detail in the literature (5,6). Their chemical structure is given in Figure 20.5.

Table 20.3: UV-absorbers (6,7)

Compound
2-(2'-Hydroxyphenyl)benzotriazoles
6-*tert*-Butyl-2-(5-chloro-2H-benzotriazol-2-yl)-4-methylphenol (Uvinul® 3026) 2,4-Di-*tert*-butyl-6-(5-chloro-2H-benzotriazol-2-yl)-phenol (Uvinul® 3027) 2-(2'-Hydroxy-5'-methylphenyl)-benzotriazole
Cyanoacrylates
Ethyl-2-cyano-3,3-diphenylacrylate (Uvinul® 3035) (2-Ethylhexyl)-2-cyano-3,3-diphenylacrylate (Uvinul® 3038)
2-Hydroxybenzophenones
2-Hydroxy-4-octyloxybenzophenon (Uvinul® 3008)
Esters of substituted and unsubstituted benzoic acids
4-*tert*-Butylphenyl salicylate Phenyl salicylate Octylphenyl salicylate Dibenzoyl resorcinol Bis(4-*tert*-butylbenzoyl)resorcinol Benzoyl resorcinol 2,4-Di-*tert*-butylphenyl 3,5-di-*tert*-butyl-4-hydroxybenzoate hexadecyl-3,5-di-*tert*-butyl-4-hydroxybenzoate
Sterically hindered amines
N,N'-Bisformyl-N,N'-bis-(2,2,6,6-tetramethyl-4-piperidinyl)-hexamethylenediamine (Uvinul® 4050 H) Bis(2,2,6,6-tetramethyl-4-piperidyl)sebacate (Uvinul® 4077 H) Bis(2,2,6,6-tetramethyl-4-piperidyl)succinate Bis(1,2,2,6,6-pentamethyl-4-piperidyl)sebacate Bis(1-octyloxy-2,2,6,6-tetramethyl-4-piperidyl)sebacate Bis(1,2,2,6,6-pentamethyl-4-piperidyl) *n*-butyl-3,5-di-*tert*-butyl-4-hydroxybenzylmalonate
Oxamides
4,4'-Dioctyloxyoxanilide 2,2'-Diethoxyoxanilide 2,2'-Dioctyloxy-5,5'-di-*tert*-butoxanilide 2,2'-Didodecyloxy-5,5'-di-*tert*-butoxanilide 2-Ethoxy-2'-ethyloxanilide N,N'-bis(3-dimethylaminopropyl)oxamide

References

1. L.K. Massey, *The Effect of UV Light and Weather on Plastics and Elastomers*, Plastics Design Library, William Andrew Publishing, Norwich, N.Y, 2nd edition, 2006.
2. H. Zweifel, ed., *Plastics Additives Handbook*, Hanser Publishers, Munich, 5th edition, 2001.
3. R. Chen and D.R. Tyler, Origin of tensile stress-induced rate increases in the photochemical degradation of polymers, *Macromolecules*, 37(14): 5430–5436, July 2004.
4. D.R. Tyler, Mechanistic aspects of the effects of stress on the rates of photochemical degradation reactions in polymers, *J. Macromol. Sci.-Polym. Rev*, C44(4):351–388, November 2004.
5. Uvinul® light stabilizers, Technical Information EVP 004605, BASF, Ludwigshafen, 2005. http://www2.basf.us/additives/pdfs/uvinul_grades_4605e.pdf.
6. Uvinul Lichtschutzmittel für Kunststoffe, Technische Information EVP 004605, BASF, Ludwigshafen, 2007. http://www.performancechemicals.basf.com/ev-wcms-in/internet/de_DE/function/conversions:/publish/upload/EV/EV1/products/pl/ti-range/evp004605d.pdf.
7. P. Piccinelli, M. Vitali, A. Landuzzi, G. Da Roit, P. Carrozza, M. Grob, and N. Lelli, Permanent surface modifiers, US Patent 7 408 077, assigned to Ciba Specialty Chemicals Corp. (Tarrytown, NY), August 5, 2008.

21

Blowing Agents

Blowing agents have been used to obtain foamable compositions of a variety of thermoplastic materials (1).

21.1 Blowing Agents

Blowing agents work by expanding a thermoplastic resin to produce a cellular thermoplastic structure having far less density than the resin from which the foam is made. Bubbles of gas form around nucleation sites and are expanded by heat or reduced pressure or by a chemical reaction in which a gas is evolved.

A nucleation site is a small particle or conglomerate of small particles that promotes the formation of a gas bubble in the resin. Additives may be incorporated into the resin to promote nucleation for a particular blowing agent and, consequently, achieve a more uniform pore distribution.

During service time, the blowing agent in the cells is replaced with air. Diffusivity of the blowing agent out of the cells relative to air coming into the cells impacts the stability of the foam and may even collapse the cells.. Additives may be incorporated into the resin and process conditions may be adjusted to assist in controlling the diffusivity of the blowing agent, agent, in order to promote foam stability, and limit the collapse of the foam to acceptable levels (2).

Blowing agents may be physical blowing agents or chemical blowing agents (3).

Table 21.1: Physical Blowing Agents and Boiling Points (3)

Compound	Bp./[°C]
n-Pentane	36
Cyclopentane	49
n-Hexane	69
Cyclohexane	81
n-Heptane	98
Toluene	111
Dichloromethane	40
Trichloromethane	61
Trichloroethylene	87
Tetrachloromethane	77
1,2-Dichloroethane	83
1,1,2-Trichlorotrifluorethane	48
Methanol	65
Ethanol	78
2-Propanol	83
Diethyl ether	35
Diisopropyl ether	69
Acetone	56
Methyl ethyl ketone	80

21.1.1 Physical Blowing Agents

Suitable blowing agents include conventional hydrocarbon or fluorocarbon physical blowing agents. The preferred hydrocarbon blowing agents are aliphatic hydrocarbons, especially those having 4–7 carbon atoms. Physical blowing agents and and their boiling points are summarized in Table 21.1.

Fluorocarbon physical blowing agents include CCl_3F, CCl_2F_2, $CHClF_2$, and $CClF_2-CClF_2$. These are commercially available as FREON 11, FREON 12, FREON 22 and FREON 114. Other halogenated hydrocarbon physical blowing agents may include methylene chloride, chloroform, CCl_4, etc. In addition, compressed gases (e.g., carbon dioxide and nitrogen) can be used as physical blowing agents.

Physical blowing agents may be incorporated already at the stage of polymerization, for example, in a styrene monomer. Thus, eventually, the blowing agent is entrapped in the polymerized bead (4).

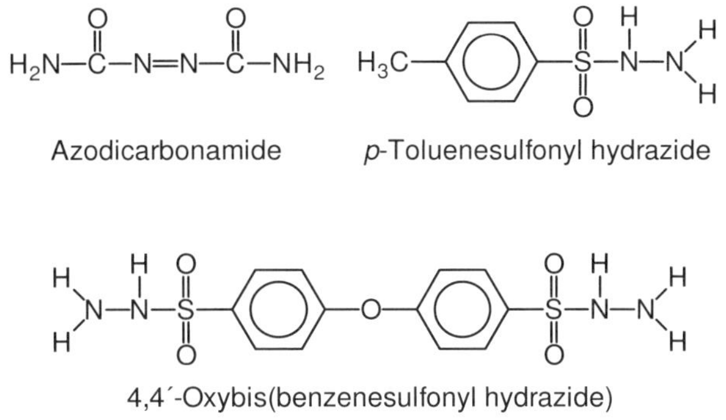

Azodicarbonamide

p-Toluenesulfonyl hydrazide

4,4′-Oxybis(benzenesulfonyl hydrazide)

Figure 21.1: Chemical Blowing Agents

Table 21.2: Chemical Blowing Agents (1,3,5)

Compound	Decomposition/[°C]
Sodium bicarbonate	130–180
Azodicarbonamide (1,1′-azobisformamide)	205–215
N,N′-Dinitroso-pentamethylenetetramine	190–210
p-Toluenesulfonyl hydrazide	110–120
4,4′-Oxybis(benzenesulfonyl hydrazide)	157–160
p-Toluenesulfonylsemicarbazide	228–235
5-Phenyltetrazole	240–250
Diisopropylhydrazodicarboxylate	
5-Phenyl-3,6-dihydro-1,3,4-oxadiazin-2-one	

21.1.2 Chemical Blowing Agents

In contrast to physical blowing agents, chemical blowing agents act by a chemical decomposition reaction in that volatile gases are produced. Chemical blowing agents are summarized and in Figure 21.1 and in Table 21.2.

A widely used blowing agent is azodicarbonamide (6). Azodicarbonamide decomposes according to the mechanism shown in Figure 21.2.

N,N′-Dinitroso-pentamethylenetetramine exhibits an interesting mechanism of decomposition, which is shown in Figure 21.3. It rearranges thermally into hexamethylenetetramine, which has the famous adamantane structure with the nitrogen and formaldehyde

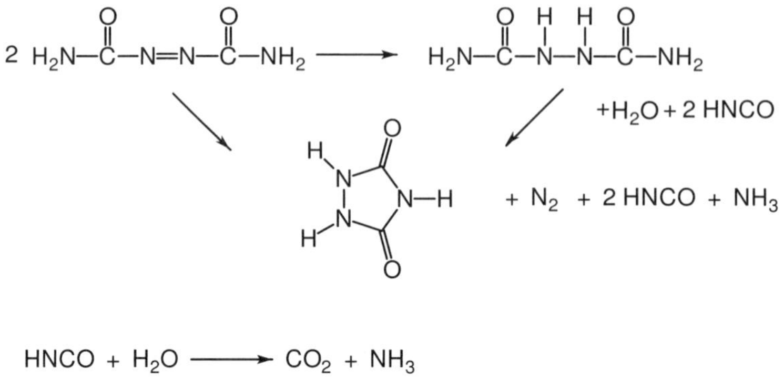

$$2\ H_2N-\overset{O}{\overset{\|}{C}}-N=N-\overset{O}{\overset{\|}{C}}-NH_2 \longrightarrow H_2N-\overset{O}{\overset{\|}{C}}-\overset{H}{\overset{|}{N}}-\overset{H}{\overset{|}{N}}-\overset{O}{\overset{\|}{C}}-NH_2$$

$+H_2O + 2\ HNCO$

$+ N_2 + 2\ HNCO + NH_3$

$HNCO + H_2O \longrightarrow CO_2 + NH_3$

Figure 21.2: Decomposition of Azodicarbonamide

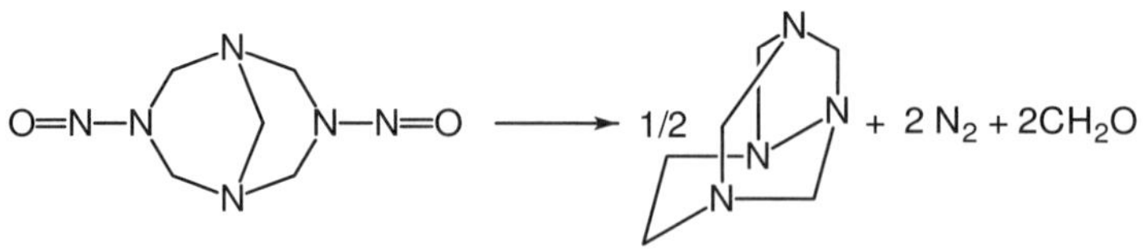

$O=N-N\cdots N-N=O \longrightarrow 1/2\ [\text{structure}] + 2\ N_2 + 2CH_2O$

Figure 21.3: Decomposition of N,N'-Dinitroso-pentamethylenetetramine

ejected. Urea can be used as a kicker. The decomposition starts already at 120°C, although the main decomposition is reached at 70°C higher. This blowing agent has been used for poly(olefin) foams (7). Another application is in polymeric microspheres in order to increase the freezing and thawing resistance of cementitious compositions (8).

A chemical blowing agent is normally uniformly dispersed in the foamable layer but is adapted upon subsequent heating to a sufficiently elevated temperature, to decompose and to free gaseous decomposition products that expand and create the foamed product. For chemically embossed sheets, the surface of a foamable polymer is printed with an ink composition containing an agent which inhibits foaming in the printed areas when the foamable polymer composition is subsequently subjected to a heat treatment. The areas which have not been printed over therefore expand normally on heating while expansion in the printed areas containing the inhibitor is suppressed, resulting in a textured surface with depressions in those areas printed with the inhibiting ink.

Based on the mechanism by which gas is freed, the compounds used for foaming polymers may be classified as chemical and physical blowing agents. Chemical blowing agents are individual solid compounds or mixtures of solid compounds that free gas as a result of chemical reactions, including thermal decomposition, or as a result of chemical reactions of chemical blowing agents. Physical blowing agents are liquid compounds that gasify as a result of physical process (evaporation, desorption) at elevated temperatures or reduced pressures.

Typical activators for azo blowing agents include acids, bases, metal organic salts, oxides, amines and urea, etc.

One critical requirement to be taken into account when selecting a chemical blowing agent is that the temperature of decomposition of the chemical blowing agents must be close to the melting point and the hardening temperature of polymer. It would be desirable to find a suitable blowing agent and activators for a metallocene poly(ethylene), which has a low melting or softening temperature, respectively, around 60°C, which is lower than 100°C (9).

The decomposition temperature range of a mixture of citric acid and sodium bicarbonate is generally at 150–210°C. The sole gaseous decomposition product is carbon dioxide (1). Neither the gaseous nor the solid decomposition products have deleterious effects on thermoplastic polymers. Furthermore, unlike numerous other blowing agents, these blowing agents do not need the presence of nucleating agents or activators in order to achieve uniformity of cells. Additionally, these blowing agents produce foams having the most uniform, fine cellular structure.

21.2 Ozone Depletion Potential

Chlorofluorocarbons (CFC)s had been used as blowing agents for rigid, closed cell insulation foams for many years because they offer outstanding fire resistance in addition to good thermal insulation, since the CFCs are non flammable. However, CFCs have been phased out because they are said to be detrimental to the ozone layer (10).

Hydrochlorofluorocarbons (HCFC)s, such as 1,1-dichloro-1-fluoroethane with low ozone depletion potential have been alternatives for CFCs. However, HCFCs are also being phased out under the Montreal Protocol. The *Montreal Protocol on Substances That Deplete the Ozone Layer* is an international treaty designed in order to protect the ozone layer by phasing out the production of a number of substances that are believed to bear an ozone depletion potential.

The next generation of foam blowing agents must have zero ozone depletion potential. For fluorochemical blowing agents, these are generally the hydrofluorocarbons (HFC)s such as 1,1,1,3,3-pentafluorobutane (HFC-365mfc). However, HFCs are typically more flammable than the CFCs or HCFCs, so that new formulations will usually require higher levels of flame retardants in order to achieve the same levels of flammability. This increased level of flame retardant creates a problem because upon burning the flame retardants increase smoke levels (10).

The use of single component fluids or azeotropic mixtures, which are mixtures that do not fractionate on boiling and evaporation, is desirable for HFC's.

Finding suitable substitutes for HCFC is complex, as any substitute should possess certain properties, such as chemical stability, low toxicity, inflammability, as well as the same efficiency of the replaced substance. An ideal substitute should not require major engineering changes in the running technology. Azeotrope-like compositions have been searched that can replace the previously used substances (11). Azeotrope-like compositions are compositions that behave close to azeotropic mixtures. In particular, mixtures of trifluoromethane, carbon dioxide, ethane, and hexafluoroethane have been investigated.

An HFC-based foam blowing agent composition has been developed. The composition contains *trans*-1,2-dichloroethylene in amounts effective to enhance the fire performance of the blown foam, as well as poly(urethane) foam compositions comprising a polyol, an isocyanate and the blowing agent composition. Preferred HFCs include 1,1,1,3,3-pentafluoropropane, and 1,1,1,2-tetrafluoroethane. Typical trans-1,2-dichloroethylene levels are from about 5 to 40% by weight, based on the total blowing agent weight (10).

21.3 Test Methods

The fire behavior of the foams can be tested with a cone calorimeter, according to standard test protocols (10, 12). The test method is used to determine the ignitability, heat release rates, mass loss rates, effective heat of combustion, and visible smoke development of materials and products.

In this test the foam specimens are ignited with a conical radiant heater, the thermal flux applied on the specimen surface being 50 kW m^{-2}. The specimens tested had a size of 100 mm by 100 mm with a thickness of 50 mm. The samples are wrapped in aluminum foil in order to have only the upper surface exposed to the radiant heater. Two specimens are used for each measurement and the results are averaged.

21.4 Special Applications

21.4.1 Poly(urethane) Foams

A rigid poly(urethane) foam or poly(isocyanurate) foam can be prepared by reacting a poly(ol) and an isocyanate in the presence of a blowing agent, a reaction catalyst, a foam stabilizer and other additives. The isocyanate is generally reacted with a premix of the polyol, the blowing agent, the reaction catalyst, the foam stabilizer and the additives in the industrial production of rigid poly(urethane) foam or poly(isocyanurate) foam.

1,1-Dichloro-1-fluoroethane was used as the blowing agent for the preparation of rigid poly(urethane) foams. However, this compound has the capability to destroy the ozone layer. So it has been decided to prohibit the use of this compound as a blowing agent.

Attention has been focused to 1,1,1,3,3-pentafluoropropane as an alternative compound, because 1,1,1,3,3-pentafluoropropane contains no chlorine atom in the molecule and thereby has no capability to destroy the ozone layer (13).

However, there are problems in using this compound as a blowing agent. It exhibits a low boiling point of 15.3°C and a low solubility in a poly(ol) composition so that a premix has a high vapor pressure and requires careful handling.

In contrast, 1,1-dichloro-1-fluoroethane has a boiling point of 32°C and a higher solubility in a poly(ol) composition owing to chlorine atoms in the molecule.

Vapor Pressure Depressants

An improved method for processing has been developed, using a vapor pressure depressant.

A vapor pressure depressant has a higher boiling point and is capable of being completely mixed with fluorocarbon. Moreover, it is in the liquid state at normal temperatures. Examples are given in Table 21.3.

Blowing Agent Enhancers

Although more environmentally acceptable blowing agents have come into use that are typically hydrohalocarbons, these are generally not as effective as those commonly used previously and there is, therefore, a continuing need for enhancements in the process of making rigid poly(urethane) foams and in the properties of the foams themselves.

Table 21.3: Vapor Pressure Depressants (13)

Compound
Dimethyl carbonate
Acetone
Methyl formate
γ-Butyrolacton
Tetrahydrofuran
Dimethoxymethane
1,3-Dioxolan
Acetonitrile
N-Methyl-2-pyrrolidone
Dimethyl sulfoxide
Sulfolane
Tris(2-chloropropyl)phosphate
Triethyl phosphate

The enhancer may comprise a low molecular weight alcohol or ether, for example di(ethylene glycol) methyl ether (14).

The effectiveness of blowing agent enhancers are shown in Table 21.4. The blowing agent enhancer with the best flow and *k*-factor is di(ethylene glycol) methyl ether.

21.4.2 Poly(imide) Foams

For poly(imide) foams suitable blowing agents include water, methanol, ethanol, acetone, 2-butoxyethanol, ethyl glycol butyl ether, ethylene glycol, halogen substituted organic compound, and ether (9).

21.4.3 Poly(ethylene) Foams

Chemically embossed metallocene polyethylene foams are used, for example in a floor covering. A blowing agent azodicarbonamide has been proposed. The blowing agent activator is selected from citric acid, oxalic acid, *p*-toluene sulfonic acid, phosphoric acid, potassium carbonate, borax, triethanol amine, zinc chloride, zinc acetate, zinc oxide, zinc stearate, barium stearate, calcium stearate, urea and poly(ethylene glycol) (6).

Table 21.4: Effectiveness of Blowing Agent Enhancers

Compound	Fill %	k^a [WK^{-1}]
No enhancer	92.5	20.6
Diethylene glycol monomethyl ether	99.2	20.6
Tripropylene glycol monobutyl ether	95.6	20.6
Propylene glycol monobutyl ether	97.7	20.9
Dipropylene glycol monopropyl ether	94.8	21.2
Propylene glycol monomethyl ether	99.0	21.2
Ethylene glycol monobutyl ether	96.1	22.1
Dipropylene glycol dimethyl ether	97.2	22.2
Dipropylene glycol monobutyl ether	96.1	21.5
Dipropylene glycol monomethyl ether	96.7	21.2
Propylene glycol monopropyl ether	97.7	21.2
Ethylene glycol phenyl ether	95.9	21.5
Tripropylene glycol monopropyl ether	95.7	21.6

^{a} Heat flow

References

1. E. Pressman, Modified flame retardant polyphenylene ether resins having improved foamability and molded articles made therefrom, US Patent 4 791 145, assigned to General Electric Company (Selkirk, NY), December 13, 1988.
2. S.T. Lee, Expandable composition and process for extruded thermoplastic foams, US Patent 5 462 974, assigned to Sealed Air Corporation (Saddle Brook, NJ), October 31, 1995.
3. J.M. Joyce and D.J. Kelley, Polyphenylene ether-alkenyl aromatic polymer blends having organobromine additives, US Patent 4 927 858, assigned to Huntsman Chemical Corporation (Salt Lake City, UT), May 22, 1990.
4. W. Burkett and M. Carnahan, Method for forming a foam product with enhanced fire resistance and product produced thereby, US Patent 6 383 608, May 7, 2002.
5. H. Zweifel, ed., *Plastics Additives Handbook*, Hanser Publishers, Munich, 5th edition, 2001.
6. L.Y.T. Yang and M. Dees, Chemically embossed metallocene polyethylene foam, US Patent 6 140 379, assigned to Armstrong World Industries, Inc. (Lancaster, PA), October 31, 2000.
7. K. Iwasa, H. Erami, N. Ueda, K. Shibayama, and J. Fukatani, Thermoplastic foam and method for production thereof, US Patent 7 173 068,

assigned to Sekisui Chemical Co., Ltd. (Osaka, JP), February 6, 2007.
8. F. Ong, Method of delivery of agents providing freezing and thawing resistance to cementitious compositions, US Patent 7 435 766, assigned to Construction Research & Technology GmbH (Trostberg, DE), October 14, 2008.
9. J.M. Vazquez, R.J. Cano, B.J. Jensen, and E.S. Weiser, Polyimide foams, US Patent 6 956 066, assigned to The United States of America as represented by the Administrator of the National Aeronautics and Space Administration (Washington, DC), October 18, 2005.
10. S.M. Galaton and C. Bertelo, Blowing agent blends, US Patent 7 144 926, assigned to Arkema Inc. (Philadelphia, PA), December 5, 2006.
11. R.R. Singh, I.R. Shankland, R.P. Robinson, H.T. Pham, R.H.P. Thomas, and P.B. Logsdon, Azeotrope-like compositions of trifluoromethane, carbon dioxide, ethane and hexafluoroethane, US Patent 5 728 315, assigned to AlliedSignal Inc. (Morristown, NJ), March 17, 1998.
12. Standard test method for heat and visible smoke release rates for materials and products using an oxygen consumption calorimeter, ASTM Standard, Book of Standards, Vol. 04.07 ASTM E 1354-08a, ASTM International, West Conshohocken, PA, 2007.
13. Y. Hibino, T. Hesaka, and N. Takada, Blowing agent, premix and process for preparing rigid polyurethane foam or polyisocyanurate foam, US Patent 7 326 362, assigned to Central Glass Company, Limited (Ube-Shi, JP), February 5, 2008.
14. J.W. Miller, Blowing agent enhancers for polyurethane foam production, US Patent 6 921 779, assigned to Air Products and Chemicals, Inc. (Allentown, PA), July 26, 2005.

22

Compatibilizers

In metallurgy, it is common that metals form mutual alloys. However, in polymer science, the situation is completely different. Miscibility of polymers is an exception (1).

Most polymers are not compatible with one another unless specific favorable interactions are present. This is because the favorable entropy of mixing is too small to overcome the unfavorable enthalpy of mixing, thus making the free energy of mixing unfavorable (2). A few pairs of immiscible and miscible polymers, respectively, are shown in Table 22.1.

The properties of polymers can be tailored to some extent by the copolymerization of two or more monomers simultaneously. In this way the monomers are mixed and fixed together on a molecular level. However, for small batches, it may be more convenient to mix homo polymers by melt blend blending. However, a serious drawback is the lack of miscibility. This issue can be circumvented to some extent by the use of proper compatibilizers. Compatibilizers are a special type of additive. Whereas certain additives do their duty simply by physical action during the whole service time, e.g., antiblocking agents, other types of additives start with a chemical reaction in the case of emergency, e.g., light stabilizers. In contrast, compatibilizers are effective by a chemical reaction already in the stage of processing.

Table 22.1: Some Pairs of Immiscible and Miscible Polymers (3, p. 692)

Polymer A	Polymer B
Immiscible Pairs	
Poly(styrene)	Poly(butadiene)
Styrene-butadiene rubber	Poly(butadiene)
Poly(styrene)	Poly(methyl methacrylate)
Poly(styrene)	Poly(dimethyl dioxane)
Poly(amide) 6,6	EPDM
Poly(amide) 6	Poly(ethylene terephthalate)
Miscible Pairs	
Poly(styrene)	Poly(phenylene ether)
Poly(styrene)	Poly(vinyl methyl ether)
Poly(vinyl chloride)	Poly(ethylene terephthalate)
Poly(methyl methacrylate)	Poly(vinylidene fluoride)
Poly(ethylene oxide)	Poly(acrylic acid)

22.1 Estimation of Compatibility

22.1.1 Glass Transition Temperature

A common measure of compatibility is the glass transition temperature. In the case of miscibility, the glass transition temperature becomes somewhat a mixture of the glass transition temperature of the individual components. The Fox equation is suitable to predict the glass transition temperature of a miscible blend (4).

In the limiting case of infinite molecular weight, the specific free volume V_f is related to the temperature T above the glass transition temperature

$$V_f = K + (\alpha_r - \alpha_g)T. \tag{22.1}$$

Here, α_r and α_g are the cubic thermal expansion coefficients in the rubbery and in the glassy state, respectively, and K is some additive constant. It was found that below the glass transition temperature basically the same relation holds. Thus it was concluded that the glass transition temperature is an iso-free-volume state. Therefore, for the glass transition temperature, $T = T_g$, the relation

$$V_f = V - V_{0,r} - \alpha_g T_g \qquad (22.2)$$

should hold. By combining Eq. 22.2 and Eq. 22.1, for the glass transition temperature the following relation is obtained:

$$T_g = \frac{K_1}{\alpha_r - \alpha_g}. \qquad (22.3)$$

In contrast, if the polymers are immiscible, the blend is composed of a superposition of the glass transition temperatures of the individual components. In the case of partial mixing, the situation becomes still more complex.

A simple experimental method to measure the glass transition temperature consists in differential scanning calorimetry.

22.1.2 Hildebrand Solubility Parameters

The Hildebrand solubility parameters δ can be predicted on the basis of the solubility of polymers in solvents with known Hildebrand solubility parameters.

The Hildebrand solubility is defined as the square root of the cohesive energy density, which is a characteristic for the intermolecular interactions in a pure liquid or solid. The solubility parameter is related to the enthalpy of mixing ΔH_m in Eq. 22.4.

$$\Delta H_m = n_s V_s \Phi_p (\delta_s - \delta_p)^2 \qquad (22.4)$$

ΔH_m	Enthalpy of mixing
n_s	mol of solvent
V_s	Molar volume of solvent
Φ_p	Volume Fraction of Polymer
δ_s	Solubility parameter of solvent
δ_p	Solubility parameter of polymer

The theory of solubility parameters was developed by Scatchard in 1931 and further refined by Hildebrand (5). Originally, the concept of solubility parameters was developed to describe the enthalpy of mixing of simple liquids. Afterwards it has been extended to polymers.

The Hildebrand parameters originate from thermodynamic considerations (6). The process of dissolving an amorphous polymer in a solvent is governed by the free energy of mixing ΔG_m, which is

$$\Delta G_m = \Delta H_m - T \Delta S_m. \quad (22.5)$$

ΔH_m is the enthalpy change on mixing, ΔS_m is the entropy change on mixing, and T is the absolute temperature. If $\Delta G_m < 0$, mixing will occur spontaneously.

The dissolution of a polymer is always accompanied with a large increase in entropy. Therefore, the sign of the enthalpy term governs the solubility behavior.

For a binary system with components (1) and (2) it was proposed that

$$\Delta H_m = \left[\left(\frac{\Delta E_1^v}{V_1} \right)^{1/2} - \left(\frac{\Delta E_2^v}{V_2} \right)^{1/2} \right]^2 V \Phi_1 \Phi_2 = [\delta_1 - \delta_2]^2 V \Phi_1 \Phi_2. \quad (22.6)$$

ΔE_i^v Enthalpy of vaporization of species i
V_i Volume of species i before mixing
Φ_i Volume fraction of species i after mixing
V Volume after mixing

According to Eq. 22.6, the solubility parameter is

$$\delta_1 = \left(\frac{\Delta E_i^v}{V_i} \right)^{1/2}.$$

Thus the solubility parameter is the square route of the energy of evaporation per unit volume of the respective compound. It is also addressed as the cohesive energy density. The physical meaning becomes more clear, if we expand the square in Eq. 22.6. Namely,

$$[\delta_1 - \delta_2]^2 = \delta_1^2 + \delta_2^2 - 2\delta_1 \delta_2.$$

Therefore, the form of this term turns out as

$$[\delta_1 - \delta_2]^2 = \left(\frac{\Delta E_1^v}{V_1} \right) + \left(\frac{\Delta E_2^v}{V_2} \right) - 2 \left(\frac{\Delta E_1^v}{V_1} \right)^{1/2} \left(\frac{\Delta E_2^v}{V_2} \right)^{1/2}. \quad (22.7)$$

During evaporation, the average distance of the molecules is increased in such a way that ideally no mutual interactions will occur. Therefore, we can interpret the energy of vaporization as that energy that is needed to overcome the forces of interaction as the molecule pass from the liquid state to the gaseous state.

We can now imagine the process of mixing two components as evaporating two pure components and then mixing them in the gas phase and condensing again into a mixture. During this process we lose the energies δ_1 and δ_2 in the first step of vaporization. On a molecular level, we assume that the δ_i reflects the energy of separation of two molecules of kind (i) in a first approximation. Thus, we deal only with the interaction of pairs of molecules. Interactions of higher order are neglected. In other words, the terms $\delta_1^2 + \delta_2^2$ reflect the energy of vaporization of two molecules of kind (1) and two molecules of kind (2), totalling 4 molecules.

During condensation we gain the energy of condensation of the mixture which is $-\delta_{1,2}$. Now the trick is to assume that $-\delta_{1,2}$ is the geometric mean of the the energy condensation of two molecules, one of kind (1) and one of kind (2), i.e.,

$$\delta_{1,2} = \delta_1^{1/2}\delta_2^{1/2}.$$

For this reason, the term $-2\delta_1^{1/2}\delta_2^{1/2}$ reflects the energy of condensation of 4 molecules. Therefore, the process of mixing can be expressed very elegantly as the square of the sum of the solubility parameters.

Eq. 22.4 is obtained from Eq. 22.6, by assuming that the volume fraction of the polymer is small, so that the volume fraction of the solvent turns to one. Moreover, the total volume approaches the volume of the solvent.

Method of Measurement with Solvents

Enthalpy of Vaporization. The energy of vaporization ΔE^v is formally at constant volume. It is related to the enthalpy of vaporization ΔH^v by

$$\Delta H^v = \Delta E^v + RT. \tag{22.8}$$

Here, R is the ideal gas constant. The term RT reflects the work of expansion at constant pressure into an ideal gas from a fluid with negligible volume. From Eq. 22.8, the solubility parameter of solvents can be determined by measuring the enthalpy of vaporization. Actually, the solubility parameter can be considered as the internal pressure of the solvent.

Activity Coefficients. The relationships between solvent solubility parameters and solvent activity coefficients γ_i can be obtained in regular solutions by the relationship

$$\begin{aligned} RT \ln \gamma_1 &= V_1 \Phi_2^2 (\delta_1 - \delta_2)^2 \\ RT \ln \gamma_2 &= V_2 \Phi_2^1 (\delta_1 - \delta_2)^2 \end{aligned} \qquad (22.9)$$

van der Waals Gas Constant. In many handbooks the van der Waals correction constants to the ideal gas law are tabulated, i.e. the internal pressure a and the co-volume b. These constants can be used to calculate the solubility parameters by the relationship

$$\delta = 1.2 \frac{a^{1/3}}{V}. \qquad (22.10)$$

Critical Pressure. The solubility parameter is also related to the critical pressure p_c, of a substance through the empirical equation

$$\delta = 1.25 p_c^{1/2}, \qquad (22.11)$$

where the critical pressure is expressed in atmospheres. Eq. 22.11 but simple to apply if critical pressure data are available.

Method of Measurement with Polymers

Solubility parameters cannot be calculated for polymers from heat of vaporization. This arises because polymers are nonvolatile. Therefore, indirect methods must be applied.

Solvent Screening. The solvency properties of a commercial polymer can be assessed by determining its solubility parameter range for each class of solvents. The solubility is classified into poor, moderate, and strong. The midpoints of these ranges may be used as single-valued quantities but ranges are more informative.

A small amount of solid polymer is placed in a test tube and the selected solvent is added in order to get a final polymer concentration of 20–50%. Optionally, the mixture can be temporarily warmed and stirred to speed up solution. The test results of solution are then evaluated at room temperature.

The resulting mixture should exhibit a single phase, free of gel particles or cloudiness. Otherwise the polymer must be termed as insoluble. Recommended solvents for these tests are summarized in Table 22.2.

Group Contribution Methods. Group contribution methods can be used according to the method of van Krevelen (7). In Table 22.3 group contributions are given.

Examples

In general, ΔG_m decreases as ΔH_m decreases. Therefore, if the two Hildebrand solubility parameters approach one another, the free energy of mixing approaches a minimum.

Terpene resins will be effective as solid solvents for an elastomer when their Hildebrand solubility parameters are close to the Hildebrand solubility parameters of the respective polymer. For example, from Table 22.4 it can be seen that pure polyterpene resins are suitable tackifiers for poly(ethylene) (PE), natural rubber, and polybutadiene polymers. Further, terpene phenol resins are suitable tackifiers for poly(vinyl acetate), poly(methyl methacrylate), and poly(ethylene terephthalate).

22.2 Compatibilizers

Some of the fundamental requirements for a compatibilizer as additive for reactive processing include (8,9):

Table 22.2: Recommended Solvents for Polymer Solubility

Solvent	δ $MPa^{1/2}$
Poorly Hydrogen Bonded	
n-Pentane	14.3
n-Heptane	15.1
Toluene	18.2
Tetrahydronaphthalene	19.4
Nitroethane	22.7
Acetonitrile	24.1
Nitromethane	26.0
Moderately Hydrogen Bonded	
Diethyl ether	15.1
n-Butyl acetate	17.4
Dibutyl phthalate	19.0
Dimethyl phthalate	21.9
Ethylene carbonate	30.1
Strongly Hydrogen Bonded	
2-Ethyl hexanol	19.4
2-Ethylbutanol	21.5
n-Pentanol	22.3
Ethanol	26.0
Methanol	29.7

Table 22.3: Group Contributions for Solubility Parameters

Group	Increment $MPa^{1/2}$
$-CH_3$	420
$=CH_2$	280
$=CH-$	140
$=C=$	0
$=CH2$	–
$=CH-$	222
$=C<$	82
$-CH=$ (aromatic)	–
$-C=$ (aromatic)	–
$-O-$	255
$-OH$	754
$-H$	140
$-S-$	460
$-SH$	–
$-F$	164
$-Cl$	471
$-Br$	614

- Optimal interfacial tension
- Sufficient and easy mixing
- Functional groups in the parent polymers
- Fast reactivity of the additive, at processing
- Enhanced adhesion between the phases in the solid state.

There is a difference between thermodynamic compatibility and technological compatibility. Thermodynamic compatibility guarantees that two polymers are miscible. However, even when thermodynamic compatibility is not established, a sufficient degree of mixing can be achieved. This is addressed as technological compatibility. This means that the blend reaches adequate useful properties. Both mechanical and chemical techniques can be used to attain technological compatibility.

22.2.1 Classification

There are two types of compatibilizers that are active by:

Table 22.4: Hildebrand Solubility Parameters δ of Solvents and Polymers (10, 11)

Solvent	δ $MPa^{1/2}$ [a]
n-Hexane	14.9
Diethyl ether	15.4
Cyclohexane	16.8
Xylene	18.2
Methyl ethyl ketone	19.3
Acetone	19.7
Ethyl alcohol	26.2
n-Butyl alcohol	28.7
Methyl alcohol	36.2
Water	48.0

Polymer	δ $MPa^{1/2}$
Poly(ethylene)	16–17
Poly(butadiene)	16–17
Poly(styrene)	17–20
Poly(vinyl acetate)	19
Poly(methyl methacrylate)	19–26
Poly(ethylene terephthalate)	19-22

[a] $MPa^{1/2} \cong 2.05 \times cal^{1/2} cm^{-3/2}$

Table 22.5: Major Classes of Compatibilizers (12)

Non Reactive	Reactive
Copolymer of A and B	A-X
Copolymer of A and C	C-X
Copolymer of C and D	A-B

- Physical principles, and
- Chemical reaction.

These major classes can be subdivided further as shown in Table 22.5. In Table 22.5, A, B are the components of the blend. The polymer C is considered as miscible with A. In the same manner, the polymer D is miscible with B. A–B are reaction products from A and B. X is a reactive group.

Non Reactive Compatibilizers

Herewith are examples that illustrate the rather abstract Table 22.5. In order to be an effective compatibilizer, the graft or block copolymer A–b–B or A–g–B should have segments of the in the range of the size of the respective homopolymers.

In the A–co–D type compatibilizers, one type of segment is identical with one component, and one component, D is not identical with B, but D is miscible with B.

Finally, in the In the C–co–D type compatibilizers, both segments are different with the components of the blend, A and B, but both C and D are miscible with A and B, respectively.

In the A–co–D type and the C–co–D type, additional thermodynamic driving forces that favor mixing may emerge (12).

Reactive Compatibilizers

Mostly in Poly(ester) type blends and poly(amide) type blends, utilize reactive compatibilizers. In Table 22.7, reactive compatibilizer systems for selected polymer pairs are listed.

Table 22.6: Non Reactive Compatibilizers for Copolymers from A and B (12, 13)

Component A	Compatibilizers	Component B	References
	Type A B		
PS	PS-g-PE	PE	(14, 15)
PS	SEBS	PE	
HIPS	PS-g-PB	PB	(16)
PS	PS-b-PMMA	PMMA	
PS	PS-g-PC	PC	
PC	PC-g-PMMA	PMMA	
PS	PS-b-PP	PP	(17)
SBS	EPDM	EPDM	(18)
	Type A D		
PPE	PS-g-PMMA	PMMA	
PPE	PS-g-PBT	PBT	
PS	PS-g-PMMA	SAN	
	Type C D		
PPO	PS-b-PMMA	SAN	(19–21)

Table 22.7: Reactive Compatibilizers for the Compatibilization of Polymer Blends

Compatibilizers	Blends
Diethylsuccinate (22), glycidyl methacrylate (23)	LDPE/PA6
Poly(ethylene-co-glycidyl methacrylate) (24)	LDPE/Starch
Maleic anhydride (25)	HDPE/HIPS
Poly(ethylene-co-glycidyl methacrylate) (26, 27)	HDPE/PET
Grafted ethylene-propylene rubber (EPM-g-MA) (28)	EPDM/PTT
Triallyl isocyanurate	Polyolefins
PP-g-(acrylic acid)	PET/PE
Ethylene glycol dimethacrylate	PBT/PP
Succinic anhydride/glycerol dimethacrylate (SGMA)	PET/PS
LDPE-g-MA (29)	PP/Clay
LDPE-g-MA (30)	LDPE/EVAl
LDPE-g-MA (31)	PP/PS
LDPE-g-MA (32)	LDPE/Starch

22.2.2 Reactive Processing

Reactive extrusion is an attractive route for polymer processing in order to carry out various reactions including polymerization, grafting, branching, and functionalization (33, 34).

Compatibilization reactions are usually performed in an extruder. Reactive processing has long been recognized as means for attaining compatibility of polymeric blending partners. Enhancement of compatibility is known to be attained by forming copolymers, using reactive processing, such as graft or block copolymers with segments capable of specific interactions or chemical reactions with the blend components that would otherwise be incompatible (33, 35).

In addition, for discontinuous work blenders have been used. Further ultrasonic assisted extrusion has been described (36, 37). The mechanical performance of polymer blends is significantly enhanced by ultrasonic assisted extrusion in comparison to conventional techniques.

22.3 Special Examples

22.3.1 Block Copolymers as Compatibilizers

One approach to compatibilize non-compatible polymers is to include in the blend a block copolymer which contains one chain segment derived from monomers compatible with one blend polymer and another chain segment derived from monomers compatible with the other blend polymer (2).

Polymer blends made from a poly(carbonate) (PC) resin and poly(isobutylene) can be compatibilized by including a minor amount of a poly(carbonate-isobutylene) block copolymer in the blend composition (38). Further, block copolymers of poly(isobutylene) and poly(dimethylsiloxane) and suggests their use as compatibilizers (39) have been suggested.

Radicalic polymerized block copolymers containing poly(isobutylene) and poly(*p*-chlorostyrene) have been described as compatibilizers (40). Similarly, copolymers containing poly(isobutylene) and *p-tert*-butylstyrene are useful compatibilizers for olefinic elastomers (2).

Macromonomers of polymeric dicyclopentadiene and polymeric ethylidene norbornene have been prepared and then terpolymerized with ethylene and propylene. This terpolymer exhibits compatibilization of ethylene-propylene rubber and butyl rubber (41).

22.3.2 Poly(olefin) Blends

A compatible blend of an olefin polymer material and an engineering thermoplastic is prepared by a process comprising (42):

1. Making an oxidized olefin polymer material that contains carboxylic acid groups,
2. Extruding a mixture of the oxidized olefin polymer material with an engineering thermoplastic, and optionally with additional virgin poly(olefin).

Examples for engineering thermoplastic materials are poly(amide)s (PA)s, PCs, poly(imide)s, and poly(ester)s. The oxidation of the poly(olefin) can be achieved by the treatment with an organic peroxide initiator, such as *tert*-butyl peroctoate (Lupersol® PMS).

Eventually, the oxidized polyolefin material will contain peroxide groups and other functionalities such as carboxylic acid, ketone, ester, and lactone groups. When the oxidized product is further processed by extrusion, the peroxide groups decompose, but the product still contains other oxygen-containing groups mentioned above. In addition, the number average and weight average molecular weight of the oxidized olefin polymer are usually much lower than that of the starting polymer due to the chain scission reactions during oxidation.

Ionomers of the oxidized olefin polymer materials can be prepared by neutralizing some or all of the carboxylic acid groups in the polymer. This can be accomplished by neutralization by a slurry process or by neutralization in the melt. Melt neutralization is preferred, since it can be done during the final step of extrusion simply by adding bases to the formulation.

Ionomeric graft copolymers with a backbone of a poly(olefin) improve the adhesion between poly(olefin)s and PAs or other polar polymers (43). The modified polymers may be used in adhesive formulations.

Functionalized poly(propylene) (PP) by radical melt grafting with monomethyl itaconate or dimethyl itaconate is a compatibilizer in poly(propylene) (PP)/poly(ethylene terephthalate) (PET) blends. Blends with compositions 15/85 and 30/70 by weight of PP and PET, prepared in a single screw extruder, revealed a very fine and uniform dispersion of the PP phase compared to the respective non-compatibilized blends.

An improved adhesion between the two phases is shown. Dimethyl itaconate as compatibilizer derived agent exhibits only a small activity to increase the impact resistance of PET in PP/PET blend. However, monomethyl itaconate is active in this respect. This finding is attributed to the hydrophilic nature of monomethyl itaconate. The tensile strength of PET in non-compatibilized blends gradually decreases with increasing content of PP. Blends containing functionalized PP exhibit, in general, higher values (44).

Poly(urethane)/poly(olefin) blend compositions are compatibilized using a zinc ionomer, based upon an ethylene/methacrylic acid/alkyl acrylate polymer, or a maleic anhydride-grafted ethylene oxide poly(olefin) elastomer (45).

Ricinoloxazoline maleate is a bifunctional compatibilizer agent. It can be grafted with the vinyl function of the maleate unit onto a poly(propylene) site by usual radical grafting, thus becoming oxazoline groups attached to the poly(propylene) chain. The oxazoline group can be reacted with the carboxyl groups of poly(butylene terephthalate) (46).

PP functionalized with glycidyl methacrylate can be used for the compatibilization of poly(propylene) and poly(butylene terephthalate) blends (47). Similar studies have been done for the grafting of glycidyl methacrylate onto linear low density poly(ethylene) (48).

22.3.3 Poly(amide) Blends

Graft copolymers of α,β-unsaturated carboxylic acids and anhydrides on a polypropylene backbone have often been used as compatibilizers for PP/PA blends. For example, a PP grafted with glycidyl and styrene moieties is a suitable compatibilizer (49). Likewise, a propylene homopolymer backbone is grafted with a styrene/methacrylic acid copolymer (PP-g-(S/MAA)) (50).

Poly(arylene ether) resins are commercially attractive materials because of their unique combination of physical, chemical, and electrical properties. Furthermore, the combination of these resins with PA resins into compatibilized blends results in additional overall properties such as chemical resistance and high strength.

As compatibilizers, dendritic poly(ester) resins are useful. This covers both dendrimers and hyperbranched polymers (51). Compatibilizers are needed, because conventional poly(arylene ether)/PA blends exhibit inadequate flow properties at the processing temperatures that are needed to minimize the thermal degradation of the resins. However, increasing the processing temperature beyond these temperatures in order to reduce viscosity of the blends results in brittle parts and surface imperfections in the final part.

Two-phase blends of PA-6 and low density poly(ethylene) (LDPE) have been prepared. Here in the course of reactive extrusion, an *in-situ* grafting of itaconic acid on the LDPE takes place. The performance of blending was tested with both neutralization and without neutralization of the acid groups of itaconic acid (52). The maximum increase with regard to the mechanical properties was achieved when magnesium hydroxide was used as a neutralizing agent.

22.3.4 Poly(carbonate) Blends

Compatible blends of PC with certain vinyl polymers have been described (35). The compatibilizer acts in the presence of a transesterification catalyst under conditions that effect transesterification. The compatibilizer is a copolymer containing hydroxyl groups, e.g., made from α-methyl-p-hydroxystyrene and methyl methacrylate. Other hydroxyl groups containing monomers may be p-hydroxystyrene and p-isopropenyl-o-cresol.

In preparing the compatible blend, the compatibilizer is reactively blended with PC and a blending partner in the presence of a transesterification catalyst. The blending partner can be a poly(olefin), styrene acrylonitrile copolymer, acrylonitrile-butadiene-styrene, poly(methyl methacrylate), or poly(styrene). Suitable transesterification catalysts include tetraphenyl phosphonium benzoate, tetraphenyl phosphonium acetate, and tetraphenyl phosphonium

Table 22.8: Effect of the Addition of a Compatibilizer to PC/SAN Blends (53)

Item	Value				
Poly(carbonate) /[%]	60	59	59	58	57
SAN /[%]	40	40	39	39	38
Compatibilizer /[%]	0	1	2	3	5
Vicat /[°C]	116	120	118	117	117
Flexural Modulus / [GPa]	2.99	2.98	3.07	3.02	3.04

phenolate.

Table 22.8 below shows the effect of the addition of a compatibilizer to PC/SAN blends (53). The addition of 1% of the compatibilizer shows an increase in the Vicat softening temperature of the resulting blend. The compatibilized blend also showed improved flexural modulus at a loading of 2% of the compatibilizer.

22.3.5 Composites of PVC and Cellulosic Materials

Wood-poly(vinyl chloride) (PVC) composites have higher flexural modulus than PVC alone. The improvement of the mechanical properties is always desirable for composite materials, and it will lead to more durable materials requiring less maintenance. The properties can be improved by the use of a compatibilizer.

PVC composites with cellulose fiber can be compatibilized using an organometallic zirconium compound. The preferred organometallic zirconium compound is cyclo[dineopentyl(diallyl)]pyrophosphato dineopentyl(diallyl)zirconate (54).

Preferred cellulose fiber-containing materials are selected from the group consisting of wood flour, wood fiber, and natural fibers, such as flax, rice hulls, sisal, jute, and kenaf.

The compatibilizing agents can be incorporated into the composite in any of several ways. For example, the agent can first be mixed with the PVC and then the natural fiber can be added to the mixture. Alternatively, the agent can first be mixed with the natural fiber and then the PVC can be added to the mixture. In another alternative, the PVC and the natural fiber are first blended together and then the agent is added to the blend. In still another alternative, all three components are mixed together simultaneously in an extruder (54).

It has been demonstrated that the addition of the compatibilizer in amounts of 0.2–0.3% of the composite weight increase both the strength and modulus by about 6–13%.

22.3.6 Packaging Applications

In packaging applications, polymers with barrier properties towards oxygen are used. These polymers include copolymers from ethylene vinyl alcohol, poly(acrylonitrile), or related copolymers, poly(vinylidene chloride), PET, or poly(ethylene naphthalate).

The performance can be still improved, if an oxygen scavenging compound, preferably a polymer, is blended. The oxygen scavenging polymer can be any organic polymer that irreversibly reacts with oxygen. In addition, the oxygen scavenging polymer should be miscible or compatible with the oxygen barrier polymer (55).

Particularly useful are polymers that bear a pendant cyclohexene structure. Thus, a suitable oxygen scavenging polymer is a copolymer from ethylene and vinyl cyclohexene.

Typically, the blend is made up with the oxygen barrier polymer as a matrix or dispersing phase, with the oxygen scavenging polymer as the dispersed phase. Because oxygen diffusion is limited through the oxygen barrier polymer matrix, oxygen scavenging by the dispersed phase oxygen scavenging polymer would become highly efficient, thus allowing enhancement of the oxygen barrier properties of the blend relative to that of the oxygen barrier polymer alone.

From both performance and processing points of view, it is typically desirable that the oxygen scavenging polymer be efficiently dispersed in the barrier polymers. Therefore, a compatibilizer may be needed to improve the miscibility or compatibility of the blend.

Preferred compatibilizers include an anhydride-modified or acid-modified poly(ethylene acrylate), poly(ethylene vinyl acetate), or PE. Another possible compatibilizer is a block copolymer of the oxygen barrier polymer or a similar polymer (55).

References

1. S. Datta and D.J. Lohse, *Polymeric Compatibilizers. Uses and Benefits in Polymer Blends*, Hanser Publishers, Munich, Vienna, New York, 1996.
2. M.F. Tse, H.C. Wang, R. Krishnamoorti, and A.H. Tsou, Polymer blend compatibilization using isobutylene-based block copolymers ii, US Patent 7 119 146, assigned to ExxonMobil Chemical Patents Inc. (Houston, TX), October 10, 2006.
3. L.H. Sperling, *Introduction to Physical Polymer Science*, Wiley Interscience, New York, 4th edition, 2005.
4. T.G. Fox, Jr. and P.J. Flory, Second-order transition temperatures and related properties of polystyrene. I. Influence of molecular weight, *J. Appl. Phys.*, 21(6):581, June 1950.
5. J.H. Hildebrand and R.L. Scott, *The solubility of nonelectrolytes*, Dover, New York, 1964, Reprint of the 3rd (1950) edition published by Reinhold.
6. E.A. Grulke, "Solubility parameter values," in J. Brandrup and E.H. Immergut, eds., *Polymer Handbook*, chapter VII, pp. 519–559. John Wiley & Sons Inc., New York, 3rd edition, 1989.
7. D.W. van Krevelen, *Properties of Polymers: Their Correlation with Chemical Structure, their Numerical Estimation and Prediction from Additive Group Contributions*, Elsevier, Amsterdam, New York, 3rd edition, 1990.
8. B.P. Livengood, B.W. Baird, and G.P. Marshall, Reactive compatibilization of polymeric components such as siloxane polymers with toner resins, US Patent 6 544 710, assigned to Lexmark International, Inc. (Lexington, KY), April 8, 2003.
9. L.A. Utracki, Compatibilization of polymer blends, *Can. J. Chem. Eng.*, 80(6):1008–1016, December 2002.
10. A.F.M. Barton, *CRC handbook of solubility parameters and other cohesion parameters*, CRC, Boca Raton, FL, 2nd edition, 1991.
11. A. Barton, *States of matter, states of mind*, Inst. of Physics Publ., Bristol, 1997.
12. F.C. Chang, "Compatibilized thermoplastic blends," in O. Olabisi, ed., *Handbook of Thermoplastics*, Vol. 41 of *Plastics engineering*, chapter 21, pp. 490–522. Marcel Dekker, New York, 1997.
13. C. Koning, M. Van Duin, C. Pagnoulle, and R. Jerome, Strategies for compatibilization of polymer blends, *Progress in Polymer Science*, 23(4): 707–757, 1998.
14. R. Fayt, R. Jerôme, and P. Teyssié, Molecular design of multicomponent polymer systems. XIV. Control of the mechanical properties of polyethylene-polystyrene blends by block copolymers, *Journal of Polymer Science Part B: Polymer Physics*, 27(4), 1989.

15. T. Appleby, F. Cser, G. Moad, E. Rizzardo, and C. Stavropoulos, Compatibilisation of polystyrene-polyolefin blends, *Polymer Bulletin*, 32(4): 479–485, April 1994.
16. A. Chirawithayaboon and S. Kiatkamjornwong, Compatibilization of high-impact density polyethylene/high-density polystyrene blends by styrene/ethylene-butylene/styrene block copolymer, *J. Appl. Polym. Sci.*, 91(2):742–755, January 2004.
17. C.J. You and D.M. Jia, Effects of styrene-ethylene/propylene diblock copolymer (SEP) on the compatibilization of PP/PS blends, *Chin. J. Polym. Sci.*, 21(4):443–446, July 2003.
18. H.J. Graf, Blend of epdm and sbr using an epdm of different origin as a compatibilizer, US Patent 6 800 691, assigned to Cooper Technology Services, LLC (Findlay, OH), October 5, 2004.
19. W.H. Jo, H.C. Kim, and D.H. Baik, Compatibilizing effect of a styrene-methyl methacrylate block copolymer on the phase behavior of poly (2, 6-dimethyl-1,4-phenylene oxide) and poly (styrene-co-acrylonitrile) blends, *Macromolecules*, 24(9):2231–2235, April 1991.
20. C. Auschra, R. Stadler, and I.G. Voigt-Martin, Poly (styrene-b-methyl methacrylate) block copolymers as compatibilizing agents in blends of poly (styrene-co-acrylonitrile) and poly (2, 9-dimethyl-1, 4-phenylene ether): 2. Influence of concentration and molecular weight of symmetric block copolymers, *Polymer*, 34:2094–2110, 1993.
21. C. Auschra and R. Stadler, Polymer alloys based on poly(2,6-dimethyl-1,4-phenylene ether) and poly(styrene-co-acrylonitrile) using poly-(styrene-b-(ethylene-co-butylene)-b-methyl methacrylate) triblock copolymers as compatibilizers, *Macromolecules*, 26(24):6364–6377, November 1993.
22. E. Passaglia, M. Aglietto, G. Ruggeri, and F. Picchioni, Formation and compatibilizing effect of the grafted copolymer in the reactive blending of 2-diethylsuccinate containing polyolefins with poly-ε-caprolactam (nylon-6), *Polym. Adv. Technol.*, 9(5):273–281, May 1998.
23. Q. Wei, D. Chionna, E. Galoppini, and M. Pracella, Functionalization of LDPE by melt grafting with glycidyl methacrylate and reactive blending with polyamide-6, *Macromol. Chem. Phys.*, 204(8):1123–1133, May 2003.
24. R.R.N. Sailaja, A.P. Reddy, and M. Chanda, Effect of epoxy functionalized compatibilizer on the mechanical properties of low-density polyethylene/plasticized tapioca starch blends, *Polym. Int.*, 50(12):1352–1359, December 2001.
25. H. Sato, S. Sasao, K. Matsukawa, Y. Kita, H. Yamaguchi, H.W. Siesler, and Y. Ozaki, Molecular structure, crystallinity, and morphology of uncompatibilized and compatibilized blends of polyethylene/nylon 12, *Macromol. Chem. Phys.*, 204(10):1351–1358, July 2003.

26. M. Pracella, L. Rolla, D. Chionna, and A. Galeski, Compatibilization and properties of poly(ethylene terephthalate)/polyethylene blends based on recycled materials, *Macromol. Chem. Phys.*, 203(10-11):1473–1485, July 2002.
27. M. Pracella and D. Chionna, Reactive compatibilization of blends of PET and PP modified by GMA grafting, *Macromol. Symp.*, 198:161–171, August 2003.
28. I. Aravind, P. Albert, C. Ranganathaiah, J.V. Kurian, and S. Thomas, Compatibilizing effect of EPM-g-MA in EPDM/poly(trimethylene terephthalate) incompatible blends, *Polymer*, 45(14):4925–4937, June 2004.
29. Y. Wang, F.B. Chen, and K.C. Wu, Twin-screw extrusion compounding of polypropylene/organoclay nanocomposites modified by maleated polypropylenes, *J. Appl. Polym. Sci.*, 93(1):100–112, July 2004.
30. C.H. Huang, J.S. Wu, C.C. Huang, and L.S. Lin, Morphological, thermal, barrier and mechanical properties of LDPE/EVOH blends in extruded blown films, *J. Polym. Res.-Taiwan*, 11(1):75–83, March 2004.
31. O.J. Danella and S. Manrich, Morphological study and compatibilizing effects on polypropylene/polystyrene blends, *Polym. Sci. Ser. A*, 45(11): 1086–1092, November 2003.
32. Y.J. Wang, W. Liu, and Z. Sun, Effects of glycerol and PE-g-MA on morphology, thermal and tensile properties of LDPE and rice starch blends, *J. Appl. Polym. Sci.*, 92(1):344–350, April 2004.
33. M. Xanthos, ed., *Reactive Extrusion, Principles and Practice*, Polymer Processing Institute Series, Hanser, München, 1992.
34. S. Al-Malaika, ed., *Reactive Modifiers for Polymers*, Blackie Academic & Professional, London, New York, 1997.
35. D.H. Bolton, P. Moulinie, D.M. Derikart, and N. Kohncke, Compatible blend of polycarbonate with vinyl (co)polymer, US Patent 7 041 732, assigned to Bayer MaterialScience LLC (Pittsburgh, PA), May 9, 2006.
36. A.I. Isayev and C.K. Hong, Novel ultrasonic process for in-situ copolymer formation and compatibilization of immiscible polymers, *Polym. Eng. Sci.*, 43(1):91–101, January 2003.
37. W. Feng and A.I. Isayev, In situ compatibilization of PP/EPDM blends during ultrasound aided extrusion, *Polymer*, 45(4):1207–1216, February 2004.
38. B. Köhler, W. Ebert, K. Horn, R. Weider, and T. Scholl, Use of polycarbonate-polyisobutylene block cocondensates as compatibilizers for polycarbonate-polyisobutylene mixtures, EP Patent 0 691 378, assigned to Bayer AG, January 10, 1996.
39. A.K. Saxena and T. Suzuki, Block copolymers of polyisobutylene and polydimethylsiloxane, US Patent 5 741 859, assigned to Dow Corning Corporation (Midland, MI), April 21, 1998.

40. G. Kaszas, J.E. Puskas, and J.P. Kennedy, Thermoplastic elastomers having isobutylene block and cyclized diene blocks, US Patent 4 910 261, assigned to Edison Polymer Innovation Corp. (EPIC) (Broadview Heights, OH), March 20, 1990.
41. M.F. Farona and J.P. Kennedy, Novel epdm-isobutylene graft copolymers, US Patent 4 599 384, assigned to The University of Akron (Akron, OH), July 8, 1986.
42. V.A. Dang, D. Dong, T.T.M. Phan, and C.Q. Song, Compatibilizing agent for engineering thermoplastic/polyolefin blend, US Patent 6 887 940, assigned to Basell Poliolefine Italia S.p.A. (Milan, IT), May 3, 2005.
43. E.C. Kelusky, Method for manufacture of modified polypropylene compositions, US Patent 5 137 975, assigned to Du Pont Canada Inc. (Mississauga, CA), August 11, 1992.
44. M. Yazdani-Pedram, H. Vega, J. Retuert, and R. Quijada, Compatibilizers based on polypropylene grafted with itaconic acid derivatives. effect on polypropylene/polyethylene terephthalate blends, *Polym. Eng. Sci.*, 43(4):960–964, April 2003.
45. J.E. Brann and S.H. Cree, Compatible thermoplastic polyurethane-polyolefin blend compositions, US Patent 6 632 879, assigned to DuPont Dow Elastomers L.L.C. (Wilmington, DE), October 14, 2003.
46. T. Vainio, G.H. Hu, M. Lambla, and J.V. Seppala, Functionalized polypropylene prepared by melt free radical grafting of low volatile oxazoline and its potential in compatibilization of PP/PBT blends, *J. Appl. Polym. Sci.*, 61(5):843–852, August 1996.
47. H. Cartier and G.H. Hu, Compatibilisation of polypropylene and poly(butylene terephthalate) blends by reactive extrusion: effects of the molecular structure of a reactive compatibiliser, *J. Mater. Sci.*, 35(8): 1985–1996, April 2000.
48. I. Pesneau, M.F. Champagne, and M.A. Huneault, Glycidyl methacrylate-grafted linear low-density polyethylene fabrication and application for polyester/polyethylene bonding, *J. Appl. Polym. Sci.*, 91 (5):3180–3191, March 2004.
49. K.T. Okamoto, K.D. Eastenson, and S.C. Guyaniyogi, Engineering resin-propylene polymer graft composition, US Patent 5 290 856, assigned to HIMONT Incorporated (Wilmington, DE), March 1, 1994.
50. A.J. DeNicola, Jr. and T.T.M. Phan, Polyolefin graft copolymer/polymide blend, US Patent 6 319 976, assigned to Montell Property Company BV (NL), November 20, 2001.
51. A. Adedeji, High flow compositions of compatibilized poly(arylene ether) polyamide blends, US Patent 6 794 450, assigned to General Electric Company (Pittsfield, MA), September 21, 2004.

52. S.S. Pesetskii, Y.M. Krivoguz, and B. Jurkowski, Structure and properties of polyamide 6 blends with low-density polyethylene grafted by itaconic acid and with neutralized carboxyl groups, *J. Appl. Polym. Sci.*, 92(3):1702–1708, May 2004.
53. D.H. Bolton, P. Moulinie, D.M. Derikart, and N. Kohncke, Compatible blend of polycarbonate with vinyl (co)polymer, US Patent 6 670 420, assigned to Bayer Corporation (Pittsburgh, PA), December 30, 2003.
54. P. Frenkel and E. Krainer, Compatibilizers for composites of PVC and cellulosic materials, US Patent 7 514 485, assigned to Chemtura Corporation (Middlebury, CT), April 7, 2009.
55. H. Yang, T.Y. Ching, and G. Cai, Compatible blend systems of oxygen barrier polymers and oxygen scavenging polymers, US Patent 7 247 390, assigned to Chevron Phillips Chemical Company, LP (The Woodlands, TX), July 24, 2007.

23

Prediction of Service Time

Final products should do their job over many years often in deleterious environments. Therefore, it is desirable to have a proper method for prediction of the service time at hand.

23.1 Accelerated Aging

Lifetime prediction studies on polymeric materials rely heavily on the use of accelerated thermal aging exposures. Most accelerated aging methods first expose the virgin material to various accelerated environments. Then the changes that occur in the material are documented. The goal is to extrapolate the accelerated results obtained in order to predict the material lifetime under ambient aging conditions (1).

23.1.1 *Cumulative Material Damage*

Cumulative material damage

The concept of cumulative damage states that a failure must be considered to be the direct result of accumulation of damage with time (1).

When a sample is subjected to a dynamic stress level r under constant frequency, temperature, moisture content, etc., the damage D can be formulated as a function of the number of load cycles n and the stress level r,

$$D = F(n,r). \tag{23.1}$$

The damage function is normalized to start at 0 and reaches 1 at failure. A particularly simple case of stress independent damage is

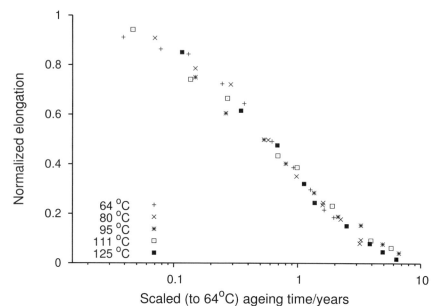

Figure 23.1: Normalized Elongation *viz.* Scaled Time at Different Aging Temperatures (1)

when $D(n,r) = D(n/n_f)$, where n_f are the number of cycles up to failure. In practice the damage functions are more complicated. In thermal experiments, instead of the number of cycles and loading, time t and temperature T may be used as parameters for the damage function.

The method has been illustrated for a nitrile rubber aged at five temperatures. The ultimate tensile elongation has been used to characterize the aging (1). It is possible to normalize and scale the data obtained at the different temperatures. A plot of the reduced elongation against the scaled time is shown in Figure 23.1.

23.1.2 Arrhenius Extrapolation

The data of aging obtained at elevated temperature are extrapolated to ambient conditions. The extrapolation is based on the law of Arrhenius (2). Thus a rate constant k is dependent on the absolute temperature T by the relation

$$k \propto \exp\left(-\frac{E_a}{RT}\right) \quad . \tag{23.2}$$

The parameter E_a is the energy of activation and R is the gas constant. Actually, the law of Arrhenius is valid for rather complex degradation processes, in that the individual rate constants may have different energies of activation.

Thus, the application of the law of Arrhenius can be trusted to be an empirical relationship that is widely usable.

However, sometimes a curvature in the logarithmic plot is observed. Most simply this curvature can be resolved in a low temperature and high temperature regime with different energy of activation E_a.

When the curvature indicates lower activation energies at lower temperatures, the service time estimated from the high temperature region would be too optimistic.

Using isothermal differential thermal analysis for the characterization of the long-term oxidation of poly(butene) and crosslinked poly(ethylene), it has been pointed out that straight line extrapolations from short-term experiments at elevated temperatures to low temperatures and long times are not possible (3). Also, oxidation induction times introduce another difficulty for interpretation of the experiments. In contrast, for isotactic poly(propylene) the application of the relationship of Arrhenius does not make problems (4).

23.1.3 Interference of Phase Transitions

Accelerated aging tests for the evaluation of antioxidants and light stabilizers, where the experiments are conducted in the melt, are considered to have little predictive value for the description of the aging process in the solid state (5).

23.2 Theory of Critical Distances

The theory of critical distances summarizes several approaches for predicting the effects of notches and other stress concentration features with regard to fracture and fatigue under a wide variety of conditions (6). The basic framework of this theory is suitable for brittle polymers, although essential modifications of the theory are necessary (7).

23.3 Monte Carlo Methods

The factors controlling the mechanical strength and mode of failure of flexible polymer fibers have been modelled with Monte Carlo methods. The model predicts available experimental data in good agreement (8).

23.4 Issues in Matrix Composites

In matrix composites, the failure is often due to creep rupture. For matrix composites, suitable models should incorporate features such as environmentally driven, statistical degradation mechanisms both in the glass fiber, and in the polymer matrix. Relevant knowledge concerning this special topic has been compiled in the literature (9).

The model should produce statistical statements on the lifetime in terms of the overall applied stress field, the overall volume of material, and boundary effects. Important input parameters are fiber packing geometry, fiber strength, matrix and interface creep exponents, rate factors in the stress-corrosion chemistry, and the applied stress level.

References

1. K.T. Gillen and M. Celina, The wear-out approach for predicting the remaining lifetime of materials, *Polymer Degradation and Stability*, 71(1): 15–30, 2000.
2. M. Celina, K.T. Gillen, and R.A. Assink, Accelerated aging and lifetime prediction: Review of non-arrhenius behaviour due to two competing processes, *Polymer Degradation and Stability*, 90(3):395–404, December 2005.
3. E. Kramer and J. Koppelmann, Measurement of oxidation stability of polyolefins by thermal analysis, *Polymer Degradation and Stability*, 16(3): 261–275, 1986.

4. G. Steiner and J. Koppelmann, Measurement of thermo-oxidative stability of isotactic polypropylene by isothermal long-term differential thermal analysis, *Polymer Degradation and Stability*, 19(4):307–314, 1987.
5. F. Gugumus, "The use of accelerated tests in the evaluation of antioxidants and light stabilizers," in G. Scott, ed., *Developments in Polymer Stabilisation*, Vol. 8, chapter 6, pp. 239–289. Elsevier Applied Science, London, 1987.
6. D. Taylor, The theory of critical distances, *Engineering Fracture Mechanics*, 2008. In Press, Corrected Proof at 2008-01-27.
7. D. Taylor, "Polymers: Brittle fracture in polymeric materials," in *The Theory of Critical Distances*, pp. 93–118. Elsevier Science Ltd, Oxford, 2007.
8. Y. Termonia, "Fracture of synthetic polymer fibers," in M. Elices and J. Llorca, eds., *Fiber Fracture*, pp. 287–302. Elsevier Science Ltd, Oxford, 2002.
9. S. Leigh Phoenix, Modeling the statistical lifetime of glass fiber/polymer matrix composites in tension, *Composite Structures*, 48(1-3):19–29, 2000.

24
Safety and Hazards

Finally, a few aspects concerning safety and health of additives are reproduced, although this is not an exhaustive by any means. The reader is advised to consult the corresponding material safety data sheets (MSDS) that can be easily found on the world wide web, e.g., in (1).

Actually, it is difficult to find review articles about health and environmental concerns of specific chemicals, although there are some sources (2,3). For a more profound understanding of the topic some monographs are recommended (4–10). In addition, the methodology of the assessment of risks has been described elsewhere (11).

24.1 Plasticizers

24.1.1 Di(2-ethylhexyl)phthalate

Di(2-ethylhexyl)phthalate (DEHP) has been reasonably anticipated to be a human carcinogen based on sufficient evidence of carcinogenicity in experimental animals (12). A detailed report is available (13).

Approximately 95% of di(2-ethylhexyl) phthalate is used as a plasticizer in poly(vinyl chloride) (PVC) resins for fabricating flexible vinyl products. Plastics may contain 1–40% of di(2-ethylhexyl) phthalate. PVC resins have been used to manufacture many products, including toys, dolls, vinyl upholstery, tablecloths, shower curtains, raincoats, garden hoses, swimming pool liners, shoes, floor tiles, polymeric coatings, components of paper and paperboard, defoaming agents, surface lubricants, disposable medical examination and surgical gloves, medical tubing, blood storage bags, flexible tubing for administering parenteral solutions, and other products.

Historically, di(2-ethylhexyl) phthalate has constituted approximately 50% of all the phthalate ester plasticizers used. However, in recent years, the use of di(2-ethylhexyl) phthalate has diminished because of health concerns. It is no longer used in plastic food packaging or baby teethers and rattles. Many toy manufacturers have discontinued its use in toys, and it is being replaced by linear phthalates and other plasticizers (12).

The primary routes of potential human exposure to di(2-ethylhexyl) phthalate are inhalation, ingestion, dermal contact, and through medical procedures. A substantial fraction of the U.S. population is exposed to measurable levels of di(2-ethylhexyl) phthalate because of its widespread use in consumer products. A high-risk segment of the population consists of individuals receiving dialysis treatments or blood transfusions from sources that have contacted di(2-ethylhexyl) phthalate-containing tubing or containers.

Workers in printing and painting occupations may also be exposed. Inhalation of aerosols or mists is the primary occupational exposure route.

Di(2-ethylhexyl) phthalate is widely distributed in the environment and has been detected in soil samples, animal and human tissues, and various forms of fish and marine life. Disposal of plastic products containing di(2-ethylhexyl) phthalate is a major source of environmental concern.

Because of its low vapor pressure, exposure to di(2-ethylhexyl) phthalate in either water or air appears to be minimal for most individuals. It is generally accepted that low levels of phthalates occur in the atmosphere throughout the United States, with higher levels near release sources. In the past, the most likely route of exposure for the general population was through contaminated food (i.e., food coming in contact with containers and wrappings containing di(2-ethylhexyl) phthalate) (12).

24.1.2 Ingestion of PVC

Four trials to evaluate the pathogenicity of soft PVC have been conducted by feeding pigs soft plastic objects. The plastic objects contained ca. 60–75% softening agents. Only in one trial did the plastic object remain in the pigs stomach. The object remained in the stomach 102 days and caused lesions of the mucous membranes. In the stomach the soft plastic became very hard and the projections of the object became pointed, resembling needles. The amount of softening agent in the object had been reduced from 69,8% to 42%. The danger, especially to children, after eating such plastics, either consciously or involuntary, is therefore demonstrated (14, 15).

Leaching of Phthalates from Infants' Toys

Humans may be exposed to phthalates from toys in many different ways. The exposure via vapors in the air is probably rather small, especially for the higher homologues as their vapor pressure is very low. The plasticizer can be transferred to the skin via direct physical contact and there is also an indirect exposure route via the food.

For small children, however, the oral exposure is probably the most likely route as they suck and chew toys. The physical massaging of the products so as there is a continuous flow of fresh saliva around the products serves as an extraction procedure for the phthalates. The main focus of this chapter will therefore be concentrated on toys for small children, especially so called teethers which are given to kids when their first teeth erupt (16).

The Laboratory of the Government Chemist, in the UK has compared leaching during static and dynamic conditions and report higher recoveries from the dynamic methods. The precision between parallel samples seems to be rather low in their study, which, according to the authors, may be explained by the release of particles from the sample. If this is the case in these experiments, it is probably a still larger problem in real life.

Several methods have been used to mimic chewing of the material. In a Dutch study, the leaching with synthetic saliva was done in an ultrasonic bath, while a study in UK tumbled the polymeric material in the saliva together with agate pellets.

Also, a report from the US shows an increase in phthalate leaching when the samples were impacted with a piston. There has also been reference to an Austrian study where actual sucking of the test material have been tested and the result indicated that amounts of phthalates migrated from PVC by the static method < the agitation method < sucking.

24.1.3 Tricresyl Phosphate

Among the aromatic phosphate esters, tricresyl phosphates with one or more *o*-cresyl groups are of toxicological importance (17).

They show a completely different clinical picture following poisoning in human beings and some animal species than that resulting from alkyl phosphate poisoning.

In the past, tricresyl phosphates have caused numerous mass poisoning incidents, for example, as plasticizers for PVC and in contaminated edible oils and machine-gun oil. The largest incidence of poisoning occurred in 1929/1930 in the United States following consumption of ginger spirit which was adulterated with tricresyl phosphate, ginger paralysis (18).

In contrast, in 2-year feed studies to rats using small doses, there was no evidence of carcinogenic activity of tricresyl phosphate (19).

More medical and biochemical details concerning the acute toxic action of tricresyl phosphates are beyond the scope of this chapter, but are documented in the literature (18, 20).

In contrast, triphenyl phosphate is not considered to be neurotoxic.

24.2 Flame Retardants

24.2.1 HET-Acid

HET-acid is also addressed as chlorendic anhydride. In particular, chlorendic anhydride is the Diels-Alder adduct of hexachlorocyclopentadiene and maleic anhydride.

Chlorendic acid and chlorendic anhydride are used as reactive flame retardants, i.e. built into the polymeric backbone, in polyester resins and as plasticizers for electrical systems and paints. Chlorendic acid is fairly persistent in soil. It has been found in landfill leachate in amounts up to 455 mg l^{-1}.

After oral and intravenous administration of radioactive labelled chlorendic acid to rats, the substance is rapidly distributed throughout the body and rapidly metabolized. Chlorendic acid has been reported to exert toxic effects on algae in concentrations of 250 mg l^{-1}.

In summary, these chemicals seem to have a low acute and subacute oral toxicity, although they are dermal, eye and respiratory irritants. From the results of long-term toxicity and carcinogenicity studies on rats and mice, it has been concluded that chlorendic acid induces tumors in rats and mice. Therefore, a carcinogenic potential is suspected. However, a full hazard assessment for humans and the environment cannot be made in view of the lack of data (21).

24.2.2 Brominated Diphenyl Ethers

Polybrominated diphenyl ethers have a large number of congeners, depending on the number and position of the bromine atoms on the two phenyl rings. Commercial brominated diphenyl ethers are produced by the bromination of diphenyl oxide. Brominated diphenyl ethers are used as flame retardants.

Since brominated diphenyl ethers are used by blending, they are referred to as additive flame retardants. These additive flame retardants are much more prone to leaching or escape from the finished polymer product than the reactive flame retardants.

Examples of polymer types, principal applications, and final products have been compiled (22). Studies upon potential risks have been conducted by industry. These studies do not reveal severe risks of the use of polybrominated diphenyl ethers.

24.3 Antifogging Agents

Basically, antifogging agents act as tensides, as explained in chapter 13. Actually, some of the antifogging agents mentioned there, as well as closely related substances, find use in related applications as tensides. These include, food industries, pharmacy, and household applications.

Of course the simultaneous use in these somehow differing fields suggests that the toxicity of these chemicals is minor. Exemplary reports can be found in the literature (23).

24.4 Other

24.4.1 Bisphenol A

Bisphenol A is used as an intermediate in the production of epoxy resins which are used in the internal coating for food and beverage cans to protect the food from direct contact with metal. It can migrate from cans with epoxy coating into foods, especially at elevated temperatures (24, 25).

In samples of canned drink products, the amount of bisphenol A was measured. Due to the particular sensitivity of the analytical method, bisphenol A was detected in samples of almost all drink products except for two tonic water soda products and one energy drink product. It is believed that quinine hydrochloride, which is commonly used as a bittering agent in tonic type drinks, may interfere with the extraction method in the analytical procedure.

The concentrations of bisphenol A in most of the drink products were generally low. 75% of the products had bisphenol A levels of less than 0.5 $\mu g\, l^{-1}$, 85% had levels less than 1 $\mu g\, l^{-1}$. The average bisphenol A level in all products was 0.57 $\mu g\, l^{-1}$. This explains why bisphenol A had not frequently been detected in canned drink products, as previously reported in the scientific literature, due to the relatively high detection limits of the methods employed (24). The results clearly indicate that exposure to bisphenol A through the consumption of canned drink products would be extremely low (25).

24.4.2 Azodicarbonamide

Azodicarbonamide is used as a blowing agent in the plastics industry. Moreover, it is used as a food additive, not – as may be suspected – as a yeast replacement, but as flour bleaching agent. However, in some countries its use as food additive is not allowed.

Azodicarbonamide is of low acute toxicity and does not cause skin, eye, or respiratory tract irritation in experimental animals. Evidence that azodicarbonamide can induce asthma in humans has been found from bronchial challenge studies with symptomatic individuals and from health evaluations of employees at workplaces where azodicarbonamide is manufactured or used. There are also indications that azodicarbonamide may induce skin sensitization (26).

Three case reports on skin sensitization have been published. For example, a male textile worker exposed to azodicarbonamide in foam ear-plugs was patch tested to discover the cause of a recurrent dermatitis of the ear (27, 28). No response was elicited with a number of standard allergens. However, the individual gave a strong positive reaction to the ear-plugs at 48 and 96 h and also to azodicarbonamide (a component of the ear-plugs) at a concentration of 1–5% in petrolatum but not at 0.1% in petrolatum. Ten control subjects patch tested with 1 and 5% azodicarbonamide in petrolatum did not respond, and the individual reported no further symptoms upon discarding the ear-plugs.

References

1. The Physical and Theoretical Chemistry Laboratory Oxford University, Chemical and other safety information, [electronic:] http://msds.chem.ox.ac.uk/, 2009.
2. INCHEM International Programe on Chemical Safety, [electronic:] http://www.inchem.org/, 2009.
3. SCORECARD The Pollution Information Site, [electronic:] http://www.scorecard.org/, 2009.
4. S. Markowitz, "Chemicals in the plastics, synthetic textiles, and rubber industries," in L. Rosenstock, M.R. Cullen, C.A. Brodkin, and C.A. Redlich, eds., *Textbook of Clinical Occupational and Environmental Medicine*, chapter 41, pp. 1011–1029. W. B. Saunders, Edinburgh, 2nd edition, 2005.
5. R. Lefaux, *Practical toxicology of plastics*, Iliffe, London, 1968, translated into English by Scripta Technica Ltd. English edition edited by Peter P. Hopf.
6. I.C. Shaw and J. Chadwick, *Principles of environmental toxicology*, Taylor & Francis, 1998.
7. P.L. Williams, R.C. James, and S.M. Roberts, eds., *Principles of Toxicology: Environmental and Industrial Applications*, Wiley, New York, 2nd edition, 2000.
8. S.F. Zakrzewski, *Principles of environmental toxicology*, Vol. 190 of *ACS monographs*, American Chemical Society, Washington, DC, 1997.
9. R.C. Dart, ed., *Medical toxicology*, Lippincott Williams & Wilkins, Philadelphia, 3rd edition, 2004.
10. J.A. Maga and A.T. Tu, *Food additive toxicology*, Dekker, New York, 1995.

11. J. Saxena and F. Fisher, eds., *Hazard Assessment of Chemicals*, Vol. 8, Academic Press, New York, NY, 1993.
12. National Toxicology Program, Di(2-ethylhexyl)phthalate, eleventh report on carcinogens, [electronic:] http://ntp.niehs.nih.gov/ntp/roc/eleventh/profiles/s087dehp.pdf, 2005.
13. Carcinogenesis bioassay of di(2-ethylhexyl)phthalate (CAS NO. 117-81-7) in F344 rats and B6C3F mice (feed study), National Toxicology Program Technical Report Series 217, U.S. Department of Health and Human Services. Public Health Service National Institutes of Health, Research Triangle Park, N.C., 1982.
14. U. Rüdt and M. Zeller, Zur Frage der Gesundheitsgefährdung durch Weich-Polyvinylchlorid nach einer per os-Aufnahme, *International Journal of Legal Medicine*, 79(2):109–114, 1977.
15. H. Altmann, W. Griem, and C. Böhme, Zur Schädigung des Verdauungstraktes beim Minischwein durch Scherzartikel aus Weich-PVC, *Bundesgesundheitsbl*, 22(15):269–274, 1979.
16. Phthalate migration from soft PVC toys and child-care articles, Opinion, EU Scientific Committee on Toxicity, Ecotoxicity and the Environment (CSTEE), Brussels, 1998. [electronic:] http://ec.europa.eu/health/ph_risk/committees/sct/documents/out12_en.pdf.
17. J. Svara, N. Weferling, and T. Hofman, "Phosphorus compounds, organic," in M. Bohnet, ed., *Phenolic Resins to Pigments, Inorganic*, Vol. 26 of *Ullmann's Encyclopedia of Industrial Chemistry*, pp. 213–243. Wiley-VCh Verlag GmbH & Co. KGaA, Weinheim, 6th edition, 2003.
18. K. Aktories, U. Förstermann, F.B. Hofmann, and K. Starke, eds., *Allgemeine und spezielle Pharmakologie und Toxikologie*, Urban & Fischer, Munich and Jena, 10th edition, 2009.
19. Toxicology and carcinogenesis studies of tricresyl phosphate (CAS NO. 1330-78-5) in F344/N rats and B6C3F, mice (gavage and feed studies), National Toxicology Program Technical Report Series 433, U.S. Department of Health and Human Services. Public Health Service National Institutes of Health, Research Triangle Park, N.C., 1994.
20. S.G. Somkuti, D.M. Lapadula, R.E. Chapin, J.C. Lamb, and M.B. Abou-Donia, Time course of the tri-o-cresyl phosphate-induced testicular lesion in F-344 rats: Enzymatic, hormonal, and sperm parameter studies, *Toxicology and Applied Pharmacology*, 89(1):64–72, June 1987.
21. G.J. van Esch, *Chlorendic Acid and Anhydride*, Vol. 17 of *Environmental Health Criteria*, World Health Organization, Geneva, 1996, [electronic:] http://www.inchem.org/documents/ehc/ehc/ehc185.htm.
22. G.J. van Esch, *Brominated Diphenyl Ethers*, Vol. 162 of *Environmental Health Criteria*, World Health Organization, Geneva, 1994, [electronic:] http://www.inchem.org/documents/ehc/ehc/ehc162.htm.

23. Joint FAO/WHO Expert Committee on Food Additives, *Polyoxyethylene (20) Sorbitan Monoesters of Lauric, Oleic, Palmitic and Stearic Acid and Triester of Stearic Acid*, number 5 in WHO Food Additives Series, World Health Organization, Geneva, 1974, [electronic:] http://www.inchem.org/documents/jecfa/jecmono/v05je47.htm.
24. X.L. Cao, J. Corriveau, and S. Popovic, Levels of bisphenol a in canned soft drink products in canadian markets, *J. Agric. Food Chem.*, 57(4):1307–1311, 2009.
25. Survey of bisphenol a in canned drink products, Publication H164-79/1-2009E-PDF, Health Canada, Ottawa, Ontario, 2009. [electronic:] http://www.hc-sc.gc.ca/fn-an/alt_formats/hpfb-dgpsa/pdf/securit/bpa_survey-enquete-can-eng.pdf.
26. R. Cary, S. Dobson, and E. Ball, eds., *Azodicarbonamide*, Vol. 16 of *Concise International Chemical Assessment Document*, World Health Organization, Geneva, 1999, [electronic:] http://www.inchem.org/documents/cicads/cicads/cicad16.htm.
27. J.L. Bonsall, Allergic contact dermatitis to azodicarbonamide, *Contact Dermatitis*, 10(1):42–49, 1984.
28. V.M. Yates and J.E. Dixon, Contact dermatitis from azodicarbonamide in earplugs, *Contact Dermatitis*, 19(2):155–156, 1988.

Index

Acronyms

ABS
  Acrylonitrile-butadiene-styrene, 98
CFC
  Chlorofluorocarbon, 202
CNT
  Carbon nanotube, 103
DOP
  Di-(2-ethylhexyl) phthalate, 9
DSC
  Differential scanning calorimetry, 121
EVA
  Ethylene vinyl acetate, 127
HCFC
  Hydrochlorofluorocarbon, 202
HDPE
  High density poly(ethylene), 114
HFC
  Hydrochlorofluorocarbon, 202
LDPE
  Low density poly(ethylene), 127, 224
LLDPE
  Linear low density poly(ethylene), 114, 127
LOI
  Limiting oxygen index, 71
NMP
  N-Methyl-2-pyrrolidone, 18
PA
  Poly(amide), 54, 74, 120, 222
PC
  Poly(carbonate), 98, 221
PE
  Poly(ethylene), 60, 96, 107, 186, 215

PET
  Poly(ethylene terephthalate), 54, 110, 223
PI
  Poly(imide), 17, 29
PLA
  Poly(lactic acid), 19, 149
PMMA
  Poly(methyl methacrylate), 98
POM
  Poly(oxymethylene), 92
PP
  Poly(propylene), 96, 108, 120, 186, 223
PPE
  Poly(phenylene ether), 79
PS
  Poly(styrene), 96
PU
  Poly(urethane), 11, 163
PVAc
  Poly(vinyl acetate), 6
PVAL
  Poly(vinyl acetal), 17
PVC
  Poly(vinyl chloride), 5, 60, 88, 96, 151, 225, 239
PVDC
  Poly(vinylidene chloride), 16
PVDF
  Poly(vinylidene fluoride), 18

Chemicals

Acetaldehyde, 54, 163
Acetic acid, 130
Acetone, 198, 205
Acetonitrile, 83, 205, 216
Acetylsalicylic acid, 166
Acetyltributyl citrate, 19
Acridine, 42
Adamantane, 199
Alumina, 32
Aluminum hydroxide, 73, 122
Aluminum 2,2'-methylene-bis(4,6-di-tert-butylphenyl) phosphate, 123
Aluminum nitride, 32
Anatase, 52
Anthranilamide, 163
Anthraquinone, 43
Antimony trioxide, 73, 81
Azodicarbonamide, 199, 244
Behenamide, 109, 142
Benzoic acid, 122
Benzoyl resorcinol, 194
Bicyclo[2.2.1]heptane-2,3-dicarboxylic acid, 121
Biphenyl, 51
4,4'-Bis((4-anilino-6-morpholino-1,3,5-triazin-2-yl)amino)stilbene-2,2'-disulfonate disodium salt, 53
1,4,-Bis(benzoxazol-2-yl)naphthalene, 53
4,4'-Bis(benzoxazol-2-yl)stilbene, 53
Bis(benzylidene)oxalyl dihydrazide, 169
2,2-Bis(3-bromo-4-hydroxyphenyl)propane, 75
2,5-Bis(5-*tert*-butyl-benzoxazol-2-yl)thiophene, 53
Bis(4-*tert*-butylbenzoyl)resorcinol, 194
2,5-Bis(4'-carbomethoxystyryl)-1,3,4-oxadiazole, 52
Bis(2,6-Dibromophenyl)methane, 75
1,2-Bis(3,5-di-*tert*-butyl-4-hydroxyhydrocinnamoyl)hydrazine, 169
N,N'-Bis(3,5-di-*tert*-butyl-4-hydroxyphenylpropionyl)hydrazine, 169
2',3-Bis[[3-[3,5-di-*tert*-butyl-4-hydroxyphenyl]propionyl]]propionohydrazide, 169
2,6-Bis(4,6-dichloronaphthyl)propane, 75
2,2-Bis(2,6-dichlorophenyl)pentane, 75
N,N'-Bis(3-dimethylaminopropyl)oxamide, 194
2,4-Bis-(3,4-dimethylbenzylidene) sorbitol, 123
2,4-Bis-(3,4-dimethyldibenzylidene) sorbitol, 125

Bis(2-ethylhexyl)phenyl phosphate, 76
Bis(2-ethylhexyl)-*p*-tolyl phosphate, 76
N,N'-Bisformyl-N,N'-bis-(2,2,6,6-tetramethyl-4-piperidinyl)-hexamethylenediamine, 194
Bis(4-Hydroxy-2,6-dichloro-3-methoxyphenyl)methane, 75
1,1-Bis-(4-iodophenyl)ethane, 75
4,4'-Bis(2-methoxystyryl)biphenyl, 53
Bis(1-octyloxy-2,2,6,6-tetramethyl-4-piperidyl)sebacate, 194
Bis(1,2,2,6,6-pentamethyl-4-piperidyl) *n*-butyl-3,5-di-*tert*-butyl-4-hydroxybenzylmalonate, 194
Bis(1,2,2,6,6-pentamethyl-4-piperidyl)sebacate, 194
Bisphenol A, 17, 83, 244
N,N'-Bis(salicyloyl)hydrazide, 167, 169
N,N'-Bis(salicyloyl)oxalyl dihydrazide, 169
N,N'-Bis(salicyloyl)thiopropionyl dihydrazide, 169
Bis(stearoyl)ethylenediamine, 90
Bis(2,2,6,6-tetramethyl-4-piperidyl)sebacate, 194
Bis(2,2,6,6-tetramethyl-4-piperidyl)succinate, 194
Boron nitride, 32
Bromelain, 148
1,3-Butadiene, 74
Butanetriol trinitrate, 20
2-Butoxyethanol, 205
n-Butyl acetate, 216
n-Butyl alcohol, 218
4-*tert*-Butylbenzoic acid, 123
Butyl benzyl phthalate, 11
6-*tert*-Butyl-2-(5-chloro-2H-benzotriazol-2-yl)-4-methylphenol, 194
1-Butyl-3-methylimidazolium hexafluorophosphate, 16
tert-Butyl peroctoate, 222
4-*tert*-Butylphenyl salicylate, 194
p-*tert*-Butylstyrene, 221
Calcium carbonate, 26, 27, 122
Calcium chloride, 161
Calcium fluoride, 122, 123
Calcium lactate, 161, 163
Calcium oxide, 78
Calcium silicate, 27
Calcium stearate, 161, 163, 205
Camphor, 5
Carbamyl poly(phosphate), 31
Cellulose acetate, 10, 19
Cellulose nitrate, 11

Index 253

Chitosan, 68
Chlorendic acid, 242, 243
Chlorendic anhydride, 77, 242
5-Chloro-2-(2,4-dichlorophenoxy)phenol, 63
2-Chloroethyl diphenyl phosphate, 76
Chloroform, 198
Chlorohexidine, 63, 68
Chlorosulfonic acid, 32
Chymotrypsin, 148
Citric acid, 79, 201, 205
Copper tetra-4-(octadecyl sulfonamido) phthalocyanine, 44
Cyclo[dineopentyl(diallyl)]pyrophosphato dineopentyl(diallyl)zirconate, 225
Cyclohexane, 198, 218
1,2-Cyclohexanedicarboxylic anhydride, 13
Cyclopentane, 198
Decabromodiphenyl oxide, 73, 75
N,N'-Diacetyladipoyl dihydrazide, 169
Diallyl phthalate, 81
4,4'-Diamino stilbene, 50
N,N-Dibenzoylhydrazine, 169
Dibenzoyl resorcinol, 194
Dibenzylidene sorbitol, 123
2,4'-Dibromobiphenyl, 75
Dibromoneopentyl glycol, 81
2,4-Di-*tert*-butyl-6-(5-chloro-2H-benzotriazol-2-yl)-phenol, 194
2,6-Di-*tert*-butyl-*p*-cresol, 182
2,6-Di-*tert*-butyl-4-methylphenol, 182
2,4-Di-*tert*-butylphenyl 3,5-di-*tert*-butyl-4-hydroxybenzoate, 194
Dibutyl phenyl phosphate, 76
Di-*i*-butyl phthalate, 11
Di-*n*-butyl phthalate, 11
Dibutyl phthalate, 18, 216
Dibutyl sebacate, 17
2,2'-Dichlorobiphenyl, 75
2,4'-Dichlorobiphenyl, 75
1,2-Dichloroethane, 32, 198
trans-1,2-Dichloroethylene, 203
1,1-Dichloro-1-fluoroethane, 202, 204
Dichloromethane, 198
4,5-Dichloro-2-*n*-octyl-4-isothiazolin-3-one, 63
Dicresylmonoxylenyl phosphate, 14
2,2'-Didodecyloxy-5,5'-di-*tert*-butoxanilide, 194

2,2'-Diethoxyoxanilide, 194
Diethylene glycol, 81
Diethylene glycol dibenzoate, 9
Di(ethylene glycol) methyl ether, 205
Diethylene glycol monomethyl ether, 206
Diethyl ether, 198, 216, 218
Di-2-ethylhexyl-l,2-cyclohexane diacid ester, 13
Di(2-ethylhexyl)phthalate, 239
2,4-Diethyl-1,5-pentanediol, 148
Diethyl phthalate, 11
Diethylsuccinate, 220
Di-*n*-hexyl phthalate, 11
9,10-Dihydro-9-oxa-10-phosphaphenanthrene-10-oxide, 83
Diisodecyl-l,2-cyclohexane diacid ester, 13
Diisodecyl phthalate, 11, 13
Diisononyl-l,2-cyclohexane diacid ester, 13
Diisononyl phthalate, 11, 13
Diisooctyl phthalate, 11
Diisopropyl ether, 198
Diisopropylhydrazodicarboxylate, 199
Diisotridecyl phthalate, 11
Dimethoxymethane, 205
2,4-Dimethoxy-6-(1'-pyrenyl)-1,3,5-triazine, 53
2,5-Dimethoxy-4-sulfonanilide phenylazo-4'-chloro-2,5-dimethoxy aceto-
    acetanilide, 44
3,4-Dimethylbenzylidene sorbitol, 122, 125
Dimethyl carbonate, 205
Dimethyl itaconate, 223
Dimethyl phthalate, 88, 216
Dimethyl sulfoxide, 205
2,2-Dinitropropyl acetal, 20
2,2-Dinitropropyl formal, 20
N,N'-Dinitroso-pentamethylenetetramine, 199
2,2'-Dioctyloxy-5,5'-di-*tert*-butoxanilide, 194
4,4'-Dioctyloxyoxanilide, 194
Di-*n*-octyl phthalate, 11
Dioctyl phthalate, 18, 88
2,4-Dioctylthiomethyl-6-methylphenol, 182
Dioxazine, 46
1,3-Dioxolan, 205
Dipentaerythrit, 158
Diphenylacetic acid, 123
Diphenylcresyl phosphate, 14

Diphenylmonoxylenyl phosphate, 14
N,N'-Diphenyloxamide, 169
Diphenyl oxide, 243
Diphenyl pentaerythritol diphosphate, 76
Dipropylene glycol dibenzoate, 9
Dipropylene glycol dimethyl ether, 206
Dipropylene glycol monobutyl ether, 206
Dipropylene glycol monomethyl ether, 206
Dipropylene glycol monopropyl ether, 206
Disodium[2.2.1]heptane bicyclodicarboxylate, 123
Distearyl phthalate, 88
Dodecyl diethylenediamine glycine, 68
Ellestadite, 27
Epichlorohydrin, 17, 100
Erucamide, 109–111, 121, 139, 142
Erucyl stearamide, 142
Ethanol, 216
2,2'-1,2-Ethenediyldi-4,1-phenylene)bisbenzoxazole, 53
2-Ethoxy-2'-ethyloxanilide, 194
Ethyl alcohol, 218
2-Ethylbutanol, 216
Ethyl-2-cyano-3,3-diphenylacrylate, 194
Ethyl diphenyl phosphate, 76
Ethylene bis stearamide, 109
Ethylene carbonate, 216
Ethylenediamine, 90
Ethylene glycol, 81, 205
Ethylene glycol dimethacrylate, 66
Ethylene glycol monobutyl ether, 206
Ethylene glycol phenyl ether, 206
Ethylene oxide, 100, 223
2-Ethyl hexanol, 216
2-Ethylhexyl)-2-cyano-3,3-diphenylacrylate, 194
2-Ethylhexyl diphenyl phosphate, 17, 76
2-Ethylhexyl di(*p*-tolyl)phosphate, 76
Ethylidene norbornene, 222
Ficin, 148
Formaldehyde, 199
Fumaric acid, 81
Glycerol monooleate, 130
Glycerol monostearate, 98
Glycidyl methacrylate, 220, 223
Graphene, 103

Graphite, 32
2,2,3,3,4,4,4-Heptafluorobutyl methacrylate, 66
n-Heptane, 198, 216
Hexabromocyclododecane, 74, 75
Hexachlorocyclopentadiene, 242
Hexachloroendomethylenetetrahydrophthalic acid, 81
Hexadecyl-3,5-di-*tert*-butyl-4-hydroxybenzoate, 194
1-Hexadecylpyridinium chloride, 63
Hexafluoroethane, 203
Hexafluoropropylene, 32
Hexahydrophthalic anhydride, 13
Hexamethylenetetramine, 199
n-Hexane, 198, 218
1-Hexyl-3-methylimidazolium dioctylsulfosuccinate, 16
1-Hexyl-3-methylimidazolium hexafluoroborate, 16
1-Hexyl-3-methylimidazolium hexafluorophosphate, 16
Hydrotalcite, 156
2-(2'-Hydroxy-5'-methylphenyl)-benzotriazole, 194
2-Hydroxy-4-octyloxybenzophenon, 194
p-Hydroxystyrene, 224
Isononanol, 15
Isophthaloyl dihydrazide, 169
p-Isopropenyl-o-cresol, 224
Isopropyl percarbonate, 66
Kaolin, 26, 122
Kaolinite, 27
Lignite, 90
Linoleamide, 109
Lithium hydroxide, 32
Lithium silicate, 133
Lysine diisocyanate, 148
Lysozyme, 67
Magnesium carbonate, 27
Magnesium hydroxide, 27, 73, 74, 224
Maleic anhydride, 81, 100, 220, 242
Melamine diphosphate, 82
Melamine pyrophosphate, 29
2-(Methacrylic acid)ethyltri-*n*-octyl phosphonium chloride, 66
Methyl alcohol, 218
p-Methylbenzylidene sorbitol, 125
4-Methyl-7-diethylaminocoumarin, 53
2-Methyl-2,5-dioxo-1-oxa-2-phospholane, 82
Methyl-di(trimethylsiloxy)silylpropyl methacrylate, 66

2,2-Methylene-bis-(4,6-di-*tert*-butylphenyl) phosphate, 125
Methylene chloride, 198
Methyl ethyl ketone, 198, 218
Methyl formate, 205
Methylisothiazolone, 63
Methyl methacrylate, 91
3-Methyl-1,5-pentanediol, 15
3-Methyl-1-phenyl-3-phosphorene-1-oxide, 146
2-Methyl-1,3-propanediol, 15
N-Methyl-2-pyrrolidone, 205
Monomethyl itaconate, 223
Monophenyldicresyl phosphate, 14
Monophenyldixylenyl phosphate, 14
Montan wax, 90
Natural rubber, 215
Neopentyl glycol, 81
Nitrocellulose, 20
Nitroethane, 216
Nitromethane, 216
Nonylphenol ethoxylate, 130
Octylphenyl salicylate, 194
Octyltin thiogylcolate, 154
Oleic acid diethanolamide, 142
Oleic amide, 142
N-Oleyl palmitamide, 142
4-(2-Oxazolyl)-phenyl-N-methyl-nitrilimine, 28
4-(2-Oxazolyl)-phenyl-*N*-methyl-nitrone, 28
4-(2-Oxazolyl)-phenyl-*N*-phenyl-nitrone, 28
4,4'-Oxybis(benzenesulfonyl hydrazide), 199
Papain, 148
Pentachlorophenyl laurate, 63
Pentaerythrit, 158
1,1,1,3,3-Pentafluorobutane, 202
1,1,1,3,3-Pentafluoropropane, 203, 204
n-Pentane, 198, 216
n-Pentanol, 216
Phenyl bis(dodecyl)phosphate, 76
Phenyl bis(neopentyl)phosphate, 76
Phenyl bis(3,5,5'-trimethylhexyl)phosphate, 76
5-Phenyl-3,6-dihydro-1,3,4-oxadiazin-2-one, 199
3-Phenyl-7-(4-methyl-6-butyloxybenzoxazole)coumarin, 53
Phenyl salicylate, 194
5-Phenyltetrazole, 199

Phthalocyanine, 42, 46
trans-1,4-Poly(butadiene), 166
3,4-Poly(ethylenedioxythiophene), 101
Potassium citrate, 161, 163
Potassium diphenylsulfone sulfonate, 75
Potassium perfluorobutane sulfonate, 75
Potassium perfluorooctane sulfonate, 75
Potassium titanate, 27
1,2-Propanediol, 81
2-Propanol, 198
Pyocyanin, 61
2,4-Pyrimidindion, 155
Pyrophyllite, 27
Quinacridone, 46
Quinoline, 42
Ricinoloxazoline maleate, 223
N-Salicylal-N'(salicyloyl)hydrazide, 169
Salicylic acid, 166
3-N-Salicyloylamino-1,2,4-triazole, 169
Sebacoyl bisphenylhydrazide, 169
Sepiolite, 27
Sericite, 27
Silicon carbide, 32
Sodium antimonate, 77
Sodium benzoate, 123
Sodium bicarbonate, 79, 199, 201
Sodium perchlorate, 100
Sodium phenyl phosphonate, 123
Sodium succinate, 123
Sorbitol, 158
Stearamide, 109
Stearic acid, 120
Stearic acid diethanolamide, 142
Stearyl behenate, 110
Stearyl(3,3-dimethyl-4-hydroxybenzyl) thioglycolate, 182
Stearyl erucamide, 109, 142
Stearyl palmitate, 110
Stearyl stearate, 92
Sulfolane, 205
Terpineol, 63
Tetrabromobisphenol A, 73, 77, 83
Tetrabromophthalic acid, 81
Tetrabromo phthalic anhydride, 77

Tetrabutyl ammonium dioctylsulfosuccinate, 16
Tetrabutyl phosphonium dioctylsulfosuccinate, 16
Tetrachloromethane, 198
Tetraethylammonium perfluorohexane sulfonate, 75
1,1,1,2-Tetrafluoroethane, 203
Tetrahydrofuran, 205
Tetrahydronaphthalene, 216
Tetrakis[methylene (3,5-di-tert-butyl-4-hydroxy)hydrocinnamate], 182
2,2,6,6-Tetramethylpiperidine, 192
Tetraphenyl phosphonium acetate, 224
Tetraphenyl phosphonium benzoate, 224
Tetraphenyl phosphonium phenolate, 225
4-(2-Thiazolyl)-phenyl-N-methyl-nitrilimine, 28
4-(2-Thiazolyl)-phenyl-N-methyl-nitrone, 28
4-(2-Thiazolyl)-phenyl-N-phenyl-nitrone, 28
3,3'-Thiodipropionic acid, 167
3,3'-Thiodipropionylhydrazide, 167
Thiosalicylic acid, 166
Thymol, 63
Titanium diboride, 31
Titanium dioxide, 115, 122, 123
Tocopherol, 178
Toluene, 198, 216
p-Toluene sulfonic acid, 205
p-Toluenesulfonyl hydrazide, 199
p-Toluenesulfonylsemicarbazide, 199
p-Tolyl bis(2,5,5'-trimethylhexyl)phosphate, 76
Triallyl isocyanurate, 220
1,2,6-Tribromophenol, 80
Tri-n-butylallyl phosphonium chloride, 66
Tributyl phosphate, 14
Tributyl (tetradecyl) phosphonium dodecylbenzenesulfonate, 16
Tributyl (tetradecyl) phosphonium methanesulfonate, 16
Trichloroethylene, 198
Trichloroethyl phosphate, 14
Trichloromethane, 198
N-Trichloromethylthiophthalimide, 63
1,1,2-Trichlorotrifluorethane, 198
Tricresyl phosphate, 5, 13, 14, 17, 76, 242
Triethanol amine, 205
Triethylene glycol, 18
Triethylene glycol dibenzoate, 9
Triethyl phosphate, 14, 205

Trifluoromethane, 203
Trihexyl (tetradecyl) phosphonium chloride, 16
Trilauryl phosphite, 182
Trimethylolpropane, 158
Tri(nonylphenyl)phosphate, 76
Tri-*n*-octylallyl phosphonium chloride, 66
Trioctyl phosphate, 14
Tri-*n*-octyl-3-vinylbenzyl phosphonium chloride, 66
Triphenyl phosphate, 14, 76, 242
Triphenyl phosphite, 182
Tripropylene glycol monobutyl ether, 206
Tripropylene glycol monopropyl ether, 206
Tris(aziridinyl) phosphine oxide, 76
Tris(2-chloropropyl)phosphate, 205
Tris(isopropylphenyl)phosphate, 14
1,1,3-Tris(2-methyl-4-hydroxy-5-*tert*-butylphenyl)butane, 182
Tris(nonylphenyl) phosphite, 182
Tritolyl phosphate, 76
Trixylyl phosphate, 14
Uracil, 155, 156
Urea, 201, 205
Vinoxycarbonylmethyltri-*n*-butyl phosphonium chloride, 66
Vinyl cyclohexene, 226
Wollastonite, 27
Xonotlite, 27
Zinc borate, 73, 74
Zinc pyrithione, 63
Zinc stearate, 116

General Index

Abrasion resistance, 91
Accelerated aging, 165, 233
Accelerated weathering, 52
Accelerator, 3
Acid neutralizers, 163
Acid scavenging, 161
Activators for blowing agents, 201
Adhesion promoters, 169
Adhesive compositions, 17, 29
Aeronautics, 81
Afterglow suppression, 74
Aging tests, 235
Agricultural films, 127
Air bags, 20
Air pollutants, 189
Algae, 59
Antacids, 161
Antiblocking agents, 109, 137, 143, 209
Antifogging additives, 127, 243
Antimicrobial agent, 59, 66
Antioxidants, 178
   antagonistic effect, 169
   phenolic, 158, 170
Antistatic additives, 95
Antistatic compositions, 100
Arrhenius extrapolation, 234
Asthma, 244
Automotive applications, 5, 11, 20, 79, 103
Autoxidation, 158, 173, 174
Azeotrope-like compositions, 203
Azeotropic mixtures, 202, 203
Baby teethers, 240
Bacteria, 59
Banbury mixer, 100, 116
Batteries, 18, 32
Beautifying patterns, 113
Beverage containers, 110
Beverages, 163

Binder compositions, 20
Biodegradability, 19
Biodegradation, 9, 19, 63, 146
Biodeterioration, 62
Blisters, 30
Blood transfusion, 240
Blowing agent enhancers, 205
Bordeaux mixture, 60
Breathable articles, 25, 141
Brighteners, 49
Brittleness, 18
Cables, 165
Calorimetry, 121, 203
Carbenes, 124
Carbon nanotubes, 30, 103, 124
   multiwalled, 104
Carcinogenicity, 239
Catalyst
   alkylation, 155
   blowing, 204
   carbodiimidation, 146
   redox, 174
   residues, 161, 166
   transesterification, 224
   Ziegler-Natta, 161
Cementitious compositions, 201
Ceramic forming, 17
Chain branching, 12, 71, 173
Chemical resistance, 224
Chromaticity diagram, 40
Chromatics, 37
Chromophores, 185
CIE value, 51
Clarifiers, 124
Clarity, 137
Clarity-enhancing agents, 119
Classification of Additives, 2
Coefficient of
   extinction, 188
   friction, 107, 108, 111, 138

thermal expansion, 210
Cohesive energy density, 212
Color coordinates, 41
Color index, 40
Color matching function, 38
Color matching system, 40
Color receptor, 39
Colorants, 37
Colorimeter, 51
   tristimulus, 41
Colorimetry, 52
Compatibility
   chain length, 108
   flame retardant, 83
   glass transition temperature, 210
   plasticizer, 7, 9, 18
   sizing composition, 31
   thermodynamic, 217
Compatibilization, 223
   polymer blends, 220
Compatibilizer, 209
Compostable packages, 148
Compounding additives, 25
Compressive stress
   degradation, 189
Condensation
   1,2,6-Tribromophenol, 80
   azo compounds, 41
   energy, 213
   polyphosphoric acids, 13
   water, 128
Conductance of carbon nanotubes, 103
Conductive compositions, 31
Conductive fillers, 30–32, 101, 102
Conductivity
   electric, 96
   electrical, 92
   ionic, 32
   thermal, 33
Cone calorimeter, 203
Congeners, 243
Conical radiant heater, 203

Conjugated double bonds, 50, 151
Conjunctiva, 65
Consumer goods, 95
Contact lenses, 65
Controlled release, 67
Conveyor belts, 11, 166
Copolymerization, 19, 66, 209
Cornea, 65
Corona discharge, 55
Corrosion inhibition, 74
Corrosivity, 162
Costabilizer
   epoxy, 158
   for PVC, 158
Counterions, 156
Crosslinking, 33, 81, 146, 152, 166, 185
Crystalline polymers, 6, 120
Crystallinity, 19
   migration, 108
   onset, 120
   spherulites, 119
Crystallites, 119, 120
Crystallization, 119, 120
   accelerator, 119, 124
   anisotropic, 123
   dynamic experiment, 121
   exotherms, 121
   nucleated, 121
   temperature, 122
Crystallization promoters, 120
Cumulative material damage, 233
Curing agent, 3
Defoaming agents, 239
Degradation, 186, 190, 235
   β-scission, 174
   autocatalytic, 151
   hydrolytic, 145
   of PVC, 152
   oxidative, 165, 167
   photochemical, 189
   photooxidative, 185
   PVC, 155

statistical, 236
thermal, 224
thermooxidative, 153
Degree of yellowness, 51
Dehydrochlorination, 151
Demolding agents, 108
Dendrimers, 224
Dendritic growth, 32
Depressants
vapor pressure, 204, 205
viscosity, 13
Dermatitis, 245
Detergents, 49
Dielectric constant, 6
Dimensional stability, 55
Discoloration, 49, 61
Discontinuous blenders, 221
Dispersion additives, 25, 28
Disposable articles, 25, 239
Distribution
molecular weight, 114
particle, 82, 102, 137, 158
pore, 197
spectral, 37–39
Drinking bottles, 55
Dripping, 74
Ductile-brittle transition, 147
Durometer, 7
Dust deposition, 134
Edible oils, 242
Einstein unit, 188
Elastomeric binders, 20, 21
Elastomeric compositions, 166
Elastomers, 14, 28, 29, 33, 169, 221
Electrodes
porous, 18
Electrolyte membranes, 18
Electronic excitation, 186
Electrostatic
chargeability, 6
discharge, 31, 100
dissipation rate, 100
Energy of activation, 235

Energy of vaporization, 213
Engineering thermoplastics, 79, 222
Entanglement limit, 147
Enthalpy
of crystallization, 122
of mixing, 209, 211
of vaporization, 212, 213
Entropy
of dissolution, 212
of mixing, 209, 212
Environmental protection, 19
Enzymatic degradability, 148
Exfoliated graphite, 103
Explosives, 20
External lubricant, 87
Extrusion, 66, 114
lubricants, 88, 108
melt, 54, 163
reactive, 221, 224
slipping, 87, 140
ultrasonic assisted, 221
Exudation, 55
Fibers
carbon, 30
natural, 225
polymer, 88, 236
stainless steel, 102
Fibrils, 123
Fire resistance, 202
Flame retardants, 30, 71, 202, 242
Flammability, 71, 202
Flax, 225
Flexible circuit laminates, 30
Flexible coverlay, 17
Flour bleaching, 244
Flow improvers, 87
Fluorescence, 186
Fluorescent whitening, 49
Fluorescents, 50
Fluorinated olefins, 114
Fluoroelastomer, 114, 163
Fluoropolymers, 114

Foamable compositions, 79, 197
Foaming agents, 79
Fogging resistance, 134
Friction, 108, 163
  coefficient of, 107, 108
  heterogeneous, 87
  lubricants, 96
  skin layer, 138
Functionalization, 221
Fungal spores, 60
Fungi, 59
Garden hoses, 239
Gas permeability, 17, 142
Gelling capacity, 10
Germination, 60
Ginger paralysis, 242
Glass laminates, 142
Glass transition temperature, 6, 17, 210
Gloss, 51, 111
Graphitization, 31
Greenhouses, 127
Group contribution, 215
Halides, 162
Hard radiation, 185
Hardening temperature, 201
Health evaluations, 244
Heat denaturation, 67
Heat resistance, 19, 43
Hildebrand solubility parameter, 211
Homolytic cleavage, 188
Hue, 50, 51
Human eye, 39
Hydrocarbon fouling, 34
Hydrogen donors, 175
Hydrolysis
  enzymatic, 146
Hydrolytic resistance, 148
Hydrolytic stabilizers, 146
Hydroperoxide decomposers, 175
Hysteresis, 25
Ignitability, 203

Immiscible polymers, 210, 211
Impact resistance, 223
Indentors, 7
Ingestion, 240
Inhibiting ink, 201
Injection molding, 66
Ink compositions, 201
Insulation
  cable, 11, 165, 167
  foams, 202
Interaggregate contact, 28
Intercalation, 162
Interconnected micropores, 34
Interfacial tension, 217
Interference pigments, 41
Interlayer films, 17
Intermetallic compounds, 31
Intermolecular
  distance, 6
  interactions, 211
Internal lubricant, 87
Intersystem crossing, 186, 188
Intravenous administration, 242
Intrinsic microporosity, 34
Intrinsically antistatic, 101
Intumescent flame retardants, 74
Ionic liquids, 15
Ionic surfactants, 98
Ionomers, 123, 222
Irritations
  by microorganisms, 60
  eye, 167
  respiratory tract, 244
Jute, 225
Lamellar orientation, 123
Laminated glass, 17
Leaching, 5, 8, 9, 14
Leaching resistance, 15
Light
  absorption, 49, 50
  accelerated weathering, 52
  degradation, 185
  diffraction, 119

emitting diodes, 3
greenhouses, 134
optical brighteners, 50
sensors, 37
spectral distribution, 39
stability, 12
tristimulus values, 37
Lithium batteries, 18, 32
Lubricants, 87, 96
Lubrication, 87, 139
Macromonomers, 221
Masterbatch, 31, 111, 115, 116, 130
Melt blending, 54
Melt fracture, 113, 115, 116
Melt viscosity, 6, 87
Membranes, 18, 34, 141
Metal deactivators, 165, 167, 174, 177
Metal soaps, 89, 155
Metallic flakes, 43
Microbicidal activity, 59
Microbiostatic agent, 59
Microglobules, 141
Microorganisms, 7, 59
Microporous fillers, 34
Microspheres, 201
Migration, 5, 67, 108, 116, 133, 139
Miscibility of polymers, 209, 210
Miscible polymers, 210
Moldable compositions, 79
Monte carlo methods, 236
Multilayer films, 67, 138, 139
Munsell color system, 40
Nano clays, 33
Nanocomposites, 103, 169
Nanofiller materials, 157
Neutralization, 222
Nitrenes, 124
Nitrilimines, 28
Nitrones, 28
Nonwovens, 134
Nucleating agents, 119
Nucleation, 119, 197

Nuclei, 119
Nutrients, 62, 65
Oil repellency, 134
Oil resistance, 15
Oligomeric plasticizers, 14
Optical brighteners, 49
Optical compensation, 50
Organoclays, 169
Organotin stabilizers, 158
Orthodontic adhesives, 11
Outdoor exposure, 52
Oxygen transmission rate, 16
Ozone depletion potential, 202
Peroxide initiators, 81
Pesticide delivery, 66
Phase transitions, 235
Phenolic antioxidants, 158, 170
Phosphorescence, 186
Photochemistry, 185
Photodegradation, 185, 186, 189
Photography, 17, 55
Photolysis, 189
Plasticization efficiency, 15
Polycondensation, 81
Polymeric plasticizers, 14
Polymerization, 65, 198, 221
free radical, 81
Polymers
biodegradable, 5, 19, 146
dendritic, 224
electrolyte, 32
energetic, 20
hyperbranched, 224
Porous electrodes, 18
Primary antioxidants, 174
Printing
curved-surface, 11
ink adhesion, 139
inkjet, 11
Printing transfer films, 11
Processability, 5, 34, 87, 91, 103
Propellant compositions, 20
Properties

antistatic, 67
rheological, 7
thermomechanical, 5
tribological, 92
viscoelastic, 7
Protease, 148
Protective coating, 97
PVC
  chlorinated, 91
  degradation, 155
  pathogenicity, 240
Raincoats, 239
RAL, 40
Rate of
  aging, 165
  antimicrobial release, 67
  combustion, 73
  cooling, 121
  crystallization, 120
  degradation, 19, 151, 189
  diffusion, 67, 189
  migration, 108
  oxidation, 169
Reaction
  acceleration, 191
  alkylation, 155
  autocatalytic, 145, 151
  autoxidation, 158, 174
  branching, 191
  bromination, 74, 243
  chain branching, 173
  compatibilization, 221
  corrosion, 161
  crosslinking, 185
  degradation, 185, 186, 222
  dehydrochlorination, 151
  Diels-Alder, 152
  explosion, 173
  hydrolysis, 146
  Norrish type, 189
  oxidation, 169
  photolysis, 189
  propagation, 191

pyrolysis, 31
redox, 165, 177
Refractive index, 137
Reinforcement, 26, 81, 123, 166
Relationship
  contact angle, 129
  ductile-brittle transition, 147
  law of Arrhenius, 235
  solubility of lubricants, 87
  solvent solubility parameter, 214
Respiratory irritants, 243
Rice hulls, 225
Rimar salt, 75
Rocket motors, 20
Rocket propellants, 20
Rolling resistance, 26
Safety glasses, 17
Scavengers
  acid, 157, 161
  hydrogen chloride, 155
  oxygen, 226
  radical, 174, 176, 190
Second order transition, 5
Secondary antioxidants, 177
Selective adsorption, 130
Semiconductor, 32
Semiconductors, 32
Separation
  energy of, 213
  gas, 34
  phase, 6
Shielding
  electromagnetic, 103
  methyl groups, 192
Shower curtains, 239
Shrinkable films, 16, 140
Silica coupling agents, 28
Singlet state, 186
Sisal, 225
Sizing compositions, 31
Skin sensitization, 244, 245
Slip agents, 107
Slurry process, 222

Smoke suppressant, 73
Solar irradiation, 133
Solubility parameters
   solvents and polymers, 218
Solvent compounding, 67
Solvolysis, 149
Spectral reflectance factors, 41
Spherulites, 119
Stabilization, 156
   hydrolytic, 146
   mechanism of, 154
   PVC, 161
   steric, 68
Stain resistance, 134
Standard allergens, 245
Static electricity, 95
Sterically hindered compounds, 20, 175, 192
Sterilization, 65
Strength
   tensile, 223
Stress-corrosion, 236
Subtractive color mixing, 39
Surface
   conductivity, 96
   slip additives, 107
   thermodynamics, 128
Surface active substances, 88
Surface blemishes, 157
Surface defects, 113
Surface improvers, 113
Surface modification, 25
Surface oxidation, 26
Surgical gloves, 240
Surgical implants, 66
Synergistic effects, 33, 73, 74, 76, 155, 158, 169
Tackifiers, 215
Tensides, 243
Terpolymer, 222
Textured surface, 201
Thermoforming, 162

Thermosetting resins, 3, 28, 33, 77, 102
Tint, 51
Tires, 26, 166
Toner compositions, 43
Toners, 45
Toxicity, 243, 244
   plasticizers, 8
   reproductive, 9
Toys, 11, 239–241
Transparency, 15, 43, 110, 119, 120, 127, 137, 142
Trimerization
   1,3-butadiene, 74
Triplet state, 186
Tristimulus colorimeter, 41
Tristimulus values, 37, 51
Ultrasonic
   assisted extrusion, 221
   leaching, 241
Van der Waals correction, 214
Varnishes, 40
Vibrational relaxation, 186
Vicat softening temperature, 225
Viruses, 59
Viscosity depressants, 13
Vulcanization, 25, 169
Water repellency, 134
Water resistance, 10
Water-dispersible compositions, 54
Weather resistance, 127
Wettability, 132
Whiteness index, 41
Whitening agents, 49
Wires, 165, 166
Wrappings, 240
X-Ray scattering, 121
Yellowing, 50, 55, 110
Yellowness index, 41, 51
Zeolithes, 66, 138
Zisman plot, 129